Infantry in Battle, 1733–1783

Alexander S. Burns

Helion & Company

Helion & Company Limited
Unit 8 Amherst Business Centre
Budbrooke Road
Warwick
CV34 5WE
England
Tel. 01926 499619
Email: info@helion.co.uk
Website: www.helion.co.uk
X (formerly Twitter): @Helionbooks
Facebook: @HelionBooks
Visit our blog at https://helionbooks.wordpress.com/

Published by Helion & Company 2025
Designed and typeset by Mach 3 Solutions (www.mach3solutions.co.uk)
Cover designed by Paul Hewitt, Battlefield Design (www.battlefield-design.co.uk)

Text © Alexander S. Burns 2024
Illustrations © as individually credited
Maps by George Anderson © Helion & Company 2024
Cover: The Battle of Germantown by Xavier della Gatta. Courtesy of the Museum of the American Revolution

Every reasonable effort has been made to trace copyright holders and to obtain their permission for the use of copyright material. The author and publisher apologise for any errors or omissions in this work, and would be grateful if notified of any corrections that should be incorporated in future reprints or editions of this book.

ISBN 978-1-804515-43-3

British Library Cataloguing-in-Publication Data.
A catalogue record for this book is available from the British Library.

All rights reserved. No part of this publication may be reproduced, stored in a retrieval system, or transmitted, in any form, or by any means, electronic, mechanical, photocopying, recording or otherwise, without the express written consent of Helion & Company Limited.

For details of other military history titles published by Helion & Company Limited, contact the above address, or visit our website: http://www.helion.co.uk

We always welcome receiving book proposals from prospective authors.

Contents

Preface		iv
Acknowledgements		v
Introduction		viii
1	Northern Italy, 1734	17
2	Armies, the Experience of Battle, and Tactics	38
3	Eighteenth-Century Soldiers and Their Village World	60
4	Theories of Battle and Prussomania: The Origins of the Clockwork Soldier Myth	80
5	Negotiated Authority: Inspiration, Religion, and Motivation	99
6	Obeying the Officers? Aimed Fire and Skirmishing	117
7	Disobeying the Officers: Range, Intensity, and Duration of Eighteenth-Century Firefights	137
8	Surviving Combat: Taking Cover, Running on the Battlefield, and Melee	157
9	North America, 1777 and 1781	177
10	Conclusion	197
Bibliography		204
Index		220

Preface

This project began 10 years ago as the *Kabinettskriege* blog while I was starting my masters degree in history at Ball State University. In the intervening years, I completed my masters and doctorate, and began my first teaching post as an assistant professor. For readers of *Kabinettskriege,* four of the nine chapters of the following book are significantly expanded versions of posts first written on that site, but even in those chapters you will find many more primary sources based on archival research in the UK, Germany, Canada, and the United States.

In most cases, I have tried to leave the names of places and people in their original language except where there is a well-known English replacement. In this way, Frederick the Great is not *Friedrich* the Great, (or *Frédéric,* as he might have preferred), and his brother Prince Henry of Prussia is not *Prinz Heinrich*. Less familiar names like Johann Dietrich Zander remain (sparing us John Derek Zander) and Vilhelm Armfeldt does not become William Armfeldt. Likewise, Cologne and Vienna remain as they are, and not their native *Köln* and *Wien*. In this usage, I have tried to follow the guidelines prescribed by John G. Gagliardo in *Germany under the Old Regime.*

When this book uses the term infantry, it specifically refers to the line infantry of the eighteenth century. Different armies used different terms: fusiliers, grenadiers, and musketeers were all primarily designed to act as infantry in the battle line. When referring specifically to light infantry or light troops, the book will use those terms to distinguish them from regular infantry. This book is specifically focused on line infantry: eighteenth-century light troops deserve specialized treatment, and helpfully, Professor Jim McIntyre has recently published such a volume.[1]

Following the pattern established by Christopher Duffy and Stephen Conway, this book uses the term 'military Europe' to refer to the international and transatlantic military culture created by the officer class of major European states during the eighteenth century. Military Europe stretched from St Petersburg to Lisbon, and across the Atlantic to armies of European colonial empires and their successor states, such as the United States and the emergent nations of Latin America. To a large extent, this international military culture was destroyed by the nationalism of the French Revolution and Napoleonic Wars.[2]

---

1   James R. McIntyre, *Light Troops in the Seven Years War, Irregular Warfare in Europe and North America, 1755-1763,* (Warwick: Helion & Co, 2024).
2   For a more complete discussion of Duffy's concept, see Alexander S. Burns (ed.), *The Changing Face of Old Regime Warfare: Essays in Honour of Christopher Duffy* (Warwick: Helion & Co, 2022) p.50.

Acknowledgements

Like a *Fahnenjunker* from a poor Pomeranian family who had recently arrived in Berlin, the young scholar writing a dissertation and first book naturally incurs a great many debts. Over the past six years as this project took shape, I have benefitted from the knowledge, kindness, wisdom, and charity of more individuals than I can count. My life would not be the same without their support, and certainly this book would never have been completed. The many errors that undoubtedly remain are solely mine.

Pride of place must go to Dr Andrew Bamford, who has done so much to make the world of eighteenth-century military history accessible to the general public, and has been a good friend and fellow traveller. His tireless work has brought about a renaissance of highly approachable studies on eighteenth-century warfare. Alongside Andrew, Robert Griffith took generous time to help the book through the editorial process.

My parents sparked my interest in history. They provided the bedrock for my interest in history to grow, and I will always be grateful for this gift. All of my grandparents nurtured my interest in the past during conversations with them about the turmoil of the twentieth century. Christopher Duffy provided a scholarly inspiration for my work, being able to edit his *Festschrift* before his passing was one of the great privileges of my life. Generously, he also helped set me on my way to an academic career with letters of recommendation. Together with Professor Duffy's writing, Dr Rory Muir's scholarly work provided needed inspiration.

Drs Sergei Zhuk, Daniel Ingram, and Nicole Etcheson continued the work of developing me as a scholar during my masters education at Ball State University between 2012 and 2014. The process of gaining an education begins with classroom instruction, and I have been fortunate enough to sit in many classrooms with my doctoral advisor at West Virginia University, Katherine B. Aaslestad. Her instruction, patience, endurance, and above all, her kindness, have made my educational journey a truly memorable one. She and her wonderful husband John Lambertson opened their home for memorable graduate student dinners.

I sincerely appreciate the vast assistance given to me by archival staff across Europe and North America. Librarians at West Virginia University and Franciscan University, particularly Judy McCracken and Amy Leoni, greatly assisted my work. Archivists helped me immensely at the Society of the Cincinnati Library, National Archives of the United States, Library and Archives of Canada, British Library, UK National Archives, Templer Study Centre, Kent Local History Centre, Geheimes Staatsarchiv Preußischer Kulturbesitz, Hessian State Archives in Wiesbaden, Darmstadt, and Marburg, and the Brandenburgisches Landeshauptarchiv. I received exemplary service at all of these fine institutions. Among the many archivists I have encountered in my travels, three stand out. At the Society of the Cincinnati Library, Director Ellen M. Clark was a repository of knowledge regarding eighteenth-century warfare, and constantly assisted my research from the time that I was completing my MA degree in 2013. At the National Army Museum in Chelsea, Robert Fleming assisted my work for two years running, and kept me well supplied with documents in the most pleasant archival environment imaginable. Finally, Herr Michael Scholz at the Hessisches Staatsarchiv Darmstadt was extremely helpful, and even retrieved a document outside of the normal daily pull times – truly an extraordinary gesture.

Numerous historians kindly donated their time to give feedback on my work, and their advice has considerably improved the final result. Sascha and Katrin Möbius gave extensive feedback on almost every aspect of my doctoral work, from explaining obscure nuances (and my many mistakes!) in Early Modern German, to assisting with my organizational knowledge of the Prussian Army. Dr Sascha Möbius also kindly agreed to serve as the outside reader of my dissertation before scheduling conflicts made that impossible. Dr Frederick C. Schneid, of High Point University, generously read several chapters and suggested improvements. Dr Matthew Keagle of Fort Ticonderoga generously fielded questions via phone and email. I am truly fortunate to have found a supportive scholarly home at Franciscan University, where the early morning pre-class conversations with Drs Kimberly Georgedes, James Matenaer, Phil Fitzgibbons, Matt O'Brien, and Robert Doyle remind me what a pleasure it is to live and work in a university history department. Dr Jonathan Abel provided a constant stream of knowledge on French military practice, encouragement, and sarcasm. I would also like to particularly thank my second anonymous peer-reviewer, who generously took the time to provide feedback and strengthen the manuscript.

Beyond professional historians, a number of individuals across Europe and North America greatly assisted my research. Hans Rosenburg generously provided selections of his ancestor *Major* Vilhelm Armfeldt's diary from the Pomeranian Seven Years War. Michał Tomaszewski kindly pointed me to the writings of Polish infantryman Marcin Matuszewicz during the War of Polish Succession. Tomasz Karpiński polished my frightful translation of Matuszewicz's description of a skirmish with the Russians. Mark Canady provided a wealth of knowledge on British practice. Jason Doerflein's conversations (and his complete library of Hans Bleckwenn's volumes) provided immeasurable support. Kyle Starich sent numerous primary sources on eighteenth-century Prussia. Jude Becker provided feedback on early drafts of the work and has been my constant companion via phone. Finally, my wife Noelle has patiently enduring my long archival trips overseas, as well as my incessant rambling about Frederick's Prussia, Noelle gave extensively editorial support to my manuscript. I could not have completed this project without her help.

In April of 2019 my advisor, Katherine Aaslestad, was diagnosed with terminal cancer. After a lengthy struggle, she passed away in April 2021, just a month after my dissertation defence. In those two years, Katherine was a model of strength, endurance, and humanity. Her tenacious spirit enabled her to continue working on projects for myself and other students until the very end. Along with all of her students, I will never forget her kindness. I endeavour to keep a small part of her alive in my work.

Alexander S. Burns
Steubenville, OH
August 2023

Introduction

What a farce it is, to see a regiment firing in divisions and subdivisions, alternatively to the right and left, when we know that in action, the officers think themselves happy, if, after the first discharge, they can make their men fire regularly in platoons.

John Hope, 1776[1]

Frameworks

Under a clear sky, with banners snapping in the wind, two armies slowly advance to the beating of drums. These armies wear brightly coloured uniforms, laced coats and hats, powdered wigs, and a bewildering array of accoutrements. Drilled to perfection, these men march slowly, robotically, inexorably, under the careful eye and direction of their officers, until they arrive within painfully close range of the enemy. By some imperceptible law of war, one side is chosen to fire first. They make a crisp volley, rolling down the line. A brief pause. Their opponents then return fire. With both sides having discharged their muskets, the armies then run into close combat with their bayonets. After a bloody and terrifying melee, one side emerges victorious.

This is the stereotypical image of eighteenth-century warfare as presented in film, works of fiction, and even historical texts, from the 1950s to the present day. Rarely is the gulf of historical reality and popular memory so far apart. As this book will show, each aspect of the preceding scenario ignores the weight of evidence, collected from the contemporary writings of officers and enlisted men who fought in the eighteenth century.

This is a book designed to give the reader an understanding of the actual experiences of infantrymen in battle between 1733 and 1783. During this 50-year period from the War of Polish Succession to the end of the American War of Independence, infantrymen shared a range of battle experiences, which seem shockingly modern when viewed from the twenty-first century. In the eighteenth-century Atlantic world and military Europe, infantrymen might still live in an early modern world, but their battles took on a more modern tenor. In some ways, the age between the Battle of Parma (1734) and the Battle of Eutaw Springs (1781) represents the birth of modern infantry warfare.

---

1   John Hope, *Hope's Curious and Comic Miscellaneous Works* (London: unknown, 1780) pp.175–176.

Stephen Biddle, a pre-eminent military theorist, defines modern infantry combat as follows:

> ...by 1914, firepower had become so lethal that exposed mass movement in the open had become suicidal... the modern system is a tightly interrelated complex of cover, concealment, dispersion, suppression, small-unit independent manoeuvre, and combined arms at the tactical level...[2]

For Biddle, improvements in weapons technology had only created these modern conditions by 1914. Even a century and a half before 1914, however, this book will show that we find infantrymen engaged in the tasks of 'cover, concealment, dispersion, suppression,' in their own way.

In the eighteenth century, many of these factors came together for the first time, but throughout history, soldiers had frequently behaved more flexibly on the battlefield than their weapons and equipment would suggest. Infantry at the Battle of Nördlingen in 1634 behaved in similar ways to the soldiers within this book: alternatively taking cover and firing.[3] What makes the eighteenth-century examples of this behaviour unique are their place as an early but identifiable ancestor of tactical practices that stretch through the nineteenth century into the present day. These men were indeed early *modern* soldiers.

Importantly, however, soldiers did not *always* act in the ways described by this book. There were moments in eighteenth-century warfare where soldiers marched slowly into fire with cadenced step, or even managed to fire by platoons in combat. Eighteenth-century warfare was sometimes rigid – but it was not always so. This book aims to bring the more flexible part of the story clearly into focus: illuminating the ways that infantry in battle between 1733 and 1783 were *both* flexible and rigid, astoundingly adaptable and flabbergastingly formal. These men were indeed *early modern* soldiers.

The book's title, *Infantry in Battle*, is an homage to Charles T. Lanham's 1934 treatise of the same name. In George C. Marshall's introduction to that work, he argues: 'There is much evidence to show that officers who have received the best peacetime training available find themselves surprised and confused by the difference between conditions as pictured in map problems and those they encounter on campaign. This is largely because our peacetime training in tactics tends to become increasingly theoretical.'[4]

Many of the issues which confronted Marshall and Lanham in the 1930s would have been familiar to eighteenth-century officers: there was a great gulf between the idealized world of military treatises and the chaos of the battlefield. This book seeks

---

2   Stephen Biddle, *Military Power: Explaining Victory and Defeat in Modern Battle* (Princeton: Princeton University Press, 2004) p.3.
3   Alberto Raúl Esteban Ribas, *The Battle of Nördlingen 1634: The Bloody Fight between Tercios and Brigades* (Warwick: Helion & Co, 2021) p.138. For primary source attestation, see: Don Diego de Aedo y Gallart, *Viaje del Infante Cardenal don Fernando de Austria* (Antwerp: Cnobbart, 1635) p.137.
4   Charles T. Lanham, *Infantry in Battle,* 2nd Edition (Richmond: Garrett & Massie, 1939), p.vii.

to bridge that gap by a focus on the experiences and agency of common soldiers, arguing that only by understanding both what is historically alien and what is modernly familiar, can we truly grapple with the world of eighteenth-century battle.

In what possible way could infantrymen with single-shot smoothbore flintlock weapons, wearing brightly coloured coats, be considered familiar or modern? This book will argue that these men were not mechanically drilled automata, but they instead aimed at their targets; could fight as skirmishers; often fired at will rather than in unison; fought at longer ranges despite their smoothbore weapons; could fight in loose order, frequently laid down and utilized cover on the battlefield; could run (rather than march) past obstacles, into cover, and out of the line of fire; and vastly preferred to employ firearms rather than melee weapons. Many officers were horrified by these things. They wanted the drilled mechanical automata of myth. By disobeying their officers, these men were engaged in a military reform process of their own: trying to fight in way that made sense to them and that they believed they could survive. Officers frequently disapproved of these deviations from the script.

The horrified reaction by officers points towards the second argument of the book. By prioritizing soldiers' writings of what occurred in battle, rather than officers' hypothetical texts on warfare, the historically minded gain a better window into the reality of the past. *Prescriptive* sources which give hypothetical advice are important, useful, and informative, but *descriptive* sources which focus on actual performance in battle provide the bedrock from which to understand theoretical writings. Trying to understand the infantrymen's world of battle necessitates understanding them and their writings. As a result, this book on battle contains a chapter analysing the world that these men hailed from, as well as the writings of their officers.

Although this book is centred on my previous research on the American, British, and Prussian armies, it is designed to cover military Europe broadly, so it includes the writings of Austrian, French, Italian, Polish, Russian, Spanish, and Swedish soldiers and officers. As Christopher Duffy asserted, by looking at military Europe with a broad lens, it is possible to identify what is regionally specific, and what is part of the shared European military experience in the Age of Reason. However, although readers will note that this book includes archival sources from multiple nations on both sides of the Atlantic, it is far from a definitive archival work. As a result, I look forward to this book being a starting place for young scholars who might archivally expand our understanding of the experience of battle in this period with a broader geographic range of archival sources.

Finally, what is the chronological framework for the book? Most anglophone readers with an interest in the eighteenth century will recognize the end date for the book as the signing of the Peace of Paris, ending (most) of the global conflict connected to the American War of Independence. English language readers may be less familiar with the book's starting date: the War of Polish Succession. Mid-eighteenth-century warfare is usually understood in terms of the cataclysmic wars of Frederick II of Prussia, and the wider Wars of Austrian Succession and Jenkins Ear which wracked the Atlantic World beginning in 1739. This is reflected in the focus of authors as well: biographies of Frederick appear in most decades, two English-language surveys of the War of Austrian Succession appeared in the 1990s, major treatments of the Seven Years War both as narrative surveys and collections of essays were released in the 2000s and 2010s.[5] The American War of Independence is so widely covered as to be its own subfield within eighteenth-century warfare.

By contrast, the most recent survey of the War of Polish Succession in English appeared nearly 45 years ago in 1980.[6] Leading Italian scholar Ciro Paoletti called out this imbalance in 2008, and the last 15 years have done little to change it.[7] And yet, the largest battles of the American War of Independence numbered between 30,000 and 40,000 combatants. These combats were similar in scale to the minor fighting in Sicily during the War of the Quadruple Alliance (1718–1720) but were dwarfed in scale by the clashes in Northern Italy during the War of Polish Succession, where 70,000 to 90,000 men fought. Although my American readers will doubtlessly argue that the fighting in the War of Independence was much more *significant* than the struggles of Polish Succession, this book is concerned with the *experience* of battle. Thus, taking up Ciro Paoletti's point, we begin our look at modern infantry combat in Lombardy during the 1730s.

Understanding Eighteenth-Century Armies

Anticipating criticism of this volume, some historians may argue that we have known infantry did not fight rigidly for some time. However, even among specialist historians of adjacent periods, the flexibility of infantry is still essentially unknown. Examine the coverage of this period in what will be the defining work on strategy and military history for the next decade or more: Hal Brands' *The New Makers of Modern Strategy*. In covering eighteenth-century battle, Michael V. Leggiere gives a standard mid-twentieth-century description:

> A rigid system of discipline based on fear bound the armies of the eighteenth century. As obedience was based on fear, desertion plagued Frederician armies. Consequently, every aspect of a Frederician army: tactics, marching, logistics was designed to prevent the individual from deserting. Tactically, the armies employed a linear system that emphasized thin, rigid, close-order lines in order to maximize firepower through a closely supervised, cohesive infantry attack… Besides possibly enabling desertion, woods and

---

5 Timothy Blanning, *Frederick the Great: King of Prussia* (New York: Random House, 2016); Reed Browning, *The War of Austrian Succession* (New York: St. Martin's Press, 1993); M.S. Anderson, *The War of Austrian Succession* (New York: Longman, 1995); Franz A. Szabo, *The Seven Years War in Europe* (New York: Pearson Longman, 2008); Mark Danley and Patrick Speelman, *The Seven Years War: Global Views* (Brill: Boston, 2012).
6 John L. Sutton, *The King's Honor and the King's Cardinal: The War of Polish Succession* (Lexington: University of Kentucky Press, 1980).
7 Ciro Paoletti, 'War, 1688–1812', in *A Companion to Eighteenth-Century Europe* (Chichester: Wiley Blackwell, 2008), p.476. Refreshingly, an English-language work on the 1747 Battle of Assietta has appeared in 2023. Giovanni Cerino Badone, *You Have to Die in Piedmont! The Battle of Assietta, 19 July 1747 and the War of Austrian Succession in the Alps* (Warwick: Helion & Co, 2023).

hills undermined the effectiveness of the volleys, broke linear cohesion, and limited the tactical control of the commanding general... generals preferred to move their units slowly and methodically over open terrain... The entire army advanced as a unitary organism toward the enemy in a completely uniform manner... Firefights featured long thin lines of three-deep infantry exchanging mass, unaimed volleys with the enemy.[8]

This assertion was made by a leading military historian, in the leading collection of essays on military strategy published in the early 2020s. This helpfully demonstrates that the image of linear warfare presented at the start of the book is not an 'aunt sally' or straw man arguments. Many people, even editors of leading works on military history, are out of touch with recent research on eighteenth-century warfare.

Although their work has yet to be as widely recognized as it might be, a new generation of scholars has revisited soldiers prior to the Napoleonic era. The recent scholarship of Ilya Berkovich and Katrin and Sascha Möbius represents a new approach to the field and has initiated a 'new school' of inquiry on eighteenth-century soldiers. Publishing their works since 2015, these historians have fundamentally altered our perception of common soldiers in old regime Europe, and it is worthwhile to briefly trace that historiography before discussing the average experiences of common soldiers themselves.[9] Before the turn of the twenty-first century, historians did not broadly question the idea that eighteenth-century common soldiers, in the turn of phrase used by the Duke of Wellington, represented 'the scum of the earth.'[10] This quote, read back into the eighteenth century, dominated historians' perception of the common soldiers of old regime armies. Historians assumed that soldiers – who they believed were often coerced peasants, misfits, or criminals – were not honourable or patriotic.[11] Fear of their officers, according to the formulation of Frederick II, motivated these soldiers. Andrew Wilson, a Professor of Strategy and Policy at the United States Naval War College, summarized this traditional view: 'In general, an ancien regime army was a slow and unwieldy mass of

---

8   Michael V. Leggiere, 'Napoleon and the Strategy of the Single Point,' in Hal Brands, *The New Makers of Modern Strategy* (Princeton: Princeton University Press, 2023) pp.320–321. In conversations with Professor Leggiere and his students, it is apparent to me that he is aware of the differing realities of eighteenth-century warfare, but there is a persistent pressure to portray a radical break in continuity between eighteenth-century warfare and Napoleonic warfare.
9   Ilya Berkovich, *Motivation in War: The Experience of Common Soldiers in Old Regime Europe* (Cambridge, Cambridge University Press, 2017); Katrin and Sascha Möbius, *The Psychology of Honour: Prussian Army Soldiers and the Seven Years War* (London: Bloomsburg Academic, 2019).
10 The full quote comes from Stanhope's conversations with the Duke of Wellington: 'I may say it in this room – are the very scum of the earth. People talk of their enlisting from their fine military feeling – all stuff – no such thing. Some of our men enlist from having got bastard children – some for minor offences – many more for drink… but you can hardly conceive such a set brought together, and it really is wonderful that we should have made them the fine fellows they are.' Philip Henry Stanhope, *Notes of Conversations with the Duke of Wellington, 1831–1851* (London: Murray, 1889), p.18.
11 For a representative example of this view, see William Willcox and Walter Arnstein, *The Age of Aristocracy: 1688–1830* (New York, Houghton Mifflin, 2001), p.110.

disgruntled and terrorized soldiers led by untrained and unimaginative officers.'[12] This traditional view was challenged by the 'new school' of specialized historians who have driven the research on old regime soldiers in a different direction, arguing that these soldiers were in fact motivated by honour, religion, localism, and state patriotism. In broad terms, then, the writings of the new school attempt to examine the place of common soldiers in the Old Regime with a critical and analytic sense of empathy in contrast to the dismissal of previous historiography. Soldiers appear in these studies, as individuals with stories, lives, and motivations, not simply as automata harshly controlled by their aristocratic officers.[13] Berkovich and Möbius recognize British historian Christopher Duffy, a lifelong specialist in eighteenth-century warfare, as the precursor of this new school, in addition to the writings of Timothy Blanning and Dennis Showalter.[14] English-language scholars have led the research making up the basis of the new school, and even those scholars native to, and covering topics related to German Central Europe, are increasingly publishing their findings in English.[15] My work joins with and contributes to this new school.

Negotiated Authority and Eighteenth-Century Warfare

For all the importance of the writings of the 'new school', historians must understand both soldiers and officers and their interaction in order to best understand what occurred on eighteenth-century battlefields. The officers and soldiers of eighteenth-century armies lived in a real-world tension with one another. On the battlefield, when lead was in the air, officers could order their men to perform an action, but whether or not the men actually performed it was, largely, up to them. As a result, officers, who believed they knew best, would often express frustration when their men failed to do exactly what they were instructed. From Habsburg grenadier captains in the 1730s, to George Washington leading the Continental Army, officers would order, beg, plead, cajole, and threaten the men into action. Sometimes the men would follow their leadership, especially when it was given in an inspiring way. Sometimes they would not. Scholars of the early-modern world often term this sort of power dynamic as 'negotiated', and one of the key contributions of this book is to try to understand the negotiated authority that eighteenth-century officers wielded on the battlefield.

The mass formations of eighteenth-century warfare made it very difficult to actually exert tactical control on the battlefield. Assuming no officers had been killed or wounded, a platoon of 75 men in the Prussian Army only had two officers and six non-commissioned officers.[16] An Austrian division of 128 men would only have 11

---

12   Andrew R. Wilson, 'Masters of War: History's Great Strategic Thinkers', lecture, *The Great Courses*, Naval War College, Newport, Rhode Island. 21 February 2012.
13   Katrin and Sascha Möbius, *Prussian Army Soldiers,* pp.3–5.
14   For a full historiographical treatment of Christopher Duffy's scholarly impact, see: Burns (ed.), *The Changing Face of Old Regime Warfare*.
15   Ilya Berkovich, Sascha Möbius, Christopher Duffy, Timothy Blanning, Dennis Showalter and Peter H. Wilson all fall into this trend.
16   Christopher Duffy, *The Army of Frederick the Great* (Warwick: Helion & Co, 2022), p.134.

officers and non-commissioned officers.[17] In the American War of Independence, there are accounts of entire platoons of infantry being managed by a sergeant and a corporal, and the corporal was one of the first men killed by enemy fire.[18] As a result, there could be no question of fully controlling every soldier: and according to recent research by Katrin and Sascha Möbius, men who disobeyed or fled from combat were rarely punished.[19] As a result, like many situations in eighteenth-century life, a negotiated authority existed on the battlefield, where the nominal authority of the officers absolutely depended on the active compliance of their men. In order to achieve that active compliance, officers needed to provide inspirational leadership that went beyond threats of physical punishment. That leadership provided the currency of battlefield negotiated authority in a mid-eighteenth-century infantry battalion.

Structure of the Book

This book is designed to take the reader chronologically and thematically through infantry battle during the 1733 to 1783 period. As a result, it begins with a chapter-length study of the battles of Parma and Guastalla in northern Italy during the War of Polish Succession in 1734. This is the first time that these battles have been studied in any tactical detail in the English language. Thematically, these battles also open up discussions on subjects including: the use of regular infantry as skirmishers, aimed and independent firing, soldiers taking cover against enemy fire, soldiers running on the battlefield, and the seeming lack of serious hand-to-hand combat. Although time did not allow for archival research in Paris, Turin, or Vienna, significant amounts of archival material on these battles, including letters from officers, are preserved in British diplomatic archives, so this chapter includes a large number of accounts from French-, Italian-, and German-speaking officers who were present at the battles in question. We can hope that the 300th anniversary of these battles in the 2030s will motivate younger scholars to produce a full tactical treatment of them.

Having opened with a battle study, the book then proceeds with four chapters that examine the make-up of armies; the soldiers who fought in them, how officers theorized on combat; and how soldiers, officers, monarchy, and religion interacted on the battlefield. The second chapter details the armies, formations, tactics, and experience of battle broadly across the 1733 to 1783 period. This chapter examines methods of deploying a battalion; the number of ranks that the battalion formed in; and the various types of firing by rank, platoons, and divisions that these troops employed on the drill square. The third chapter tours the village environment that these soldiers were drawn from, the ways that their family ties sustained them during

---

17 Christopher Duffy, *Instrument of War: The Austrian Army in the Seven Years War Volume 1* (Warwick: Helion & Co, 2020), pp.441–442.
18 Nathaniel Root, 'The Battle of Princeton', *Pennsylvania Magazine of History and Biography*, vol.20 (1896), pp.516–517.
19 Katrin and Sascha Möbius, *The Psychology of Honour*, pp.163–167.

military service, the correspondence of the surprising number of soldiers' letters that have survived to the present, and the ways that the early modern village community prepare soldiers to contest the authority of their officers. The fourth chapter sheds light on the phenomenon of Prussomania, and how officers theorized about the world of battle. The fifth chapter defines the concept of negotiated authority, exploring how officers used inspirational acts to gain currency with their men, and how the religious and monarchical ideologies of the Prussian Army sustained it in contrast to the robotic imaginings of the Prussomaniacs. The sixth chapter explores how soldiers and their officers actually fought, examining the topics of aimed fire and skirmishing. The seventh chapter, in some ways the heart of the book, illuminates the ways in which soldiers frequently ignored their officers' wishes when it came to the range, intensity, and duration of infantry firefights. The eighth chapter follows soldiers as they tried to survive battle, and sheds light on their use of cover, the ways that they would run – rather than march – on the battlefield, and their desire to avoid hand-to-hand combat. The final chapter, in many ways a mirror to the first, summarizes these themes in connection to two battles from the American War of Independence, Germantown, and Eutaw Springs. These battles, much like the fighting at Parma and Guastalla, demonstrate all of the key features of infantry combat that the book has explored up to this point.

The conclusion takes what we have learned into the setting of a fictitious battle: using the writings of enlisted soldiers and officers to reconstruct a fabricated infantry battle, demonstrating many of the themes of the book. In this way, readers can clearly see the view of a battle from men in the ranks, rather than the abstracted writings of military theorists or even army commanders. By providing a fictious battle that is full of historical sources and footnotes, the book concludes by giving the reader a glimpse into the infantrymen's world of battle between 1733 and 1783.

1

Northern Italy, 1734

Well, my dear mother, yesterday there was a great battle here under the cannon of Parma, which was one of the fiercest and bloodiest we have seen in a long time.[1]

'Count de la Roque' writing from the camp at Parma 30 June 1734

Few eighteenth-century wars have been as ill-used by historians as the War of Polish Succession of 1733–1735. Christopher Duffy called it 'the last of the dynastic wars in the old style', and never addressed it in any detail.[2] Peter Wilson spends eight pages on the war in *German Armies: War and German Politics*, but focuses primarily on the mobilization of the Holy Roman Empire for the conflict.[3] This is a war, according to historians, that is much less significant and interesting than the War of Austrian Succession. In many ways this view has some merit, but for the historian of battle, this war provides not only two large clashes between European armies, but also a starting point for a history of mid-eighteenth-century battle. The themes and methods of infantry battle that emerged in the two battles of Parma and Guastalla would shape debate around the theory and practice of infantry warfare among European armies for the next half-century. The fighting at Parma and Guastalla was in no way the norm of eighteenth-century infantry warfare, but it did demonstrate the variety of possibilities that eighteenth-century warfare could assume.

After the close of the War of Spanish Succession in 1715, almost two decades of relative peace reigned in Europe. There were minor clashes, like the War of the Quadruple Alliance of 1718–1720, but these were resolved relatively quickly, without large scale field battles of over 50,000 participants. Compared to the relatively titanic clashes of the War of Spanish Succession, these were more minor conflicts. Then in February of 1733, Augustus II, the Elector of Saxony and King of Poland-Lithuania died. His death caused a succession crisis, as rival candidates vied for the votes of the Polish *Sejm*, or elective legislature. Foreign powers backed their preferred candidates, and war in Poland and Europe was the result. The French, Spanish, and

---

1   Société Savoisenne D'Historie et D'Archéologie, *Mémoires et Documents* (Chambéry: Imprimerie Ménard, 1896), vol.XXXV, p.xiv.
2   Christopher Duffy, *The Military Experience in the Age of Reason* (New York: Atheneum, 1988), p.320.
3   Peter H. Wilson, *German Armies: War and German Politics* (London: University College London Press, 1998), pp.226–234.

Piedmont-Sardinians coalesced into a bloc supporting Polish nobleman and former king, Stanisław Leszczyński; the Austrians, Russians, Saxons, and minor states of the Holy Roman Empire supported the son of the dead king, the future Augustus III. The major fighting of this war came, not in Poland itself, where Russian forces swiftly seized control, but in northern Italy in the Po River valley, and along the Rhine at the Siege of Philippsburg.

Here, the rival factions headed by the French Bourbons and Austrian Habsburgs squared off with the largest armies that Europe had seen assembled since the War of Spanish Succession. Prince Eugene of Savoy, one of the great captains and a leader of the War of Spanish Succession era, fought his last campaigns in the generally quiet Rhine theatre, leading a Habsburg-Imperial army of 74,000 men against a French force of nearly 100,000.[4] That French force, led by James FitzJames, the 1st Duke of Berwick, laid siege and captured Philippsburg during May–July 1734. Berwick lost his head to cannon fire during the siege, and the marquis d'Asefeld finished the siege. Eugene disappointed the assembled military men by waging a campaign that lacked intensity and allowing the French to capture Philippsburg without fighting a major battle in trying to relieve the fortress. As these events progressed on the Rhine, another drama, with important legacies for the history of infantry battle in the eighteenth century, was unfolding in the Po River valley of northern Italy.

The French, Spanish, and Piedmont-Sardinians all hoped for a successful war in Italy. The French sought to reduce Habsburg influence in the region, while the Spanish and Sardinians desired actual gains in territory from the conflict, the Spanish being interested due to family connections. The Habsburgs, by contrast, sought to maintain the status quo. As a result, the French and Sardinian forces cooperated in northern Italy in the area of Milan and Mantua, while the Spanish focused their efforts in southern Italy. The Habsburgs, likewise, sent a field army into northern Italy. As a result, these armies fought two battles, in relatively quick succession, in June and September of 1734.

François de Franquetot, duc de Coigny (born 1670), and François-Marie, duc de Broglie (born 1671) led the French forces. Coigny was a cavalry officer in the Nine Years War (1688–1697), fought as a *lieutenant-général* at Malpaquet and Denain during the War of Spanish Succession (1701–1715), and as a result of the action at Colonro in early June 1734, was promoted to *maréchal de France*. His subordinate and almost exact contemporary, Broglie, was likewise present at Malpaquet and Denain. The French were assisted by allied Sardinian forces commanded by Charles Emmanuel III, Duca di Savoia and King of Sardinia. Charles Emmanuel (born 1701), the second son of Savoyard ruler Victor Amadeus II, had been thrust into the forefront with the death of his older brother in 1715. Ruling Piedmont-Sardinia from Turin, Charles Emmanuel had the unfortunate task of trying to safeguard the gains and royal title made by his father in the face of the comparatively giant states and militaries of France, Spain, and the Habsburgs.[5] In the War

---

4   Christopher Duffy, *A Military Life of Frederick the Great* (London: Routledge, 1985), p.15.
5   For recent work on Charles Emmanuel's army and efforts later in the century, see Giovanni Cerino Badone, *You Have to Die in Piedmont!*. For a good though dated summary of the larger period in English, see Spenser Wilkinson *The Defence of Piedmont, 1742–1748: A*

of Polish Succession, Charles Emmanuel found himself aligned the Spanish and French, against the Habsburgs. The French and Sardinian forces in the northern Italian theatre amounted to around 60,000 men, and they brought 53,000 to bear at the Battle of Parma.[6]

Habsburg and Imperial forces in Italy were led by Claude Florimond de Mercy, Friedrich Ludwig von Württemberg-Winnental, and Lothar Joseph Dominik von Königsegg-Rothenfels. Mercy (born 1666) was the senior commander, and the oldest, rashest, and nastiest of the three men. He was old enough to have served at the great siege of Vienna in 1683 and fought at Zenta (1697), Schellenberg (1704), Peterwardein (1716), Belgrade (1717), and Francavilla (1719). He was 68 years of age in 1734. Württemberg (born 1690) served as a junior officer in the Dutch army during the War of Spanish Succession, fought for the Saxons in the closing stages of the Great Northern War, and fought against the Ottomans at Peterwardein and Belgrade. Königsegg (born 1673), like the other two commanders, had experience of war against the Turks and the French in the War of Spanish Succession, particularly alongside Prince Eugene and Victor Amadeus II in Italy at the Battle of Turin in 1706. As the initial commanders of the Habsburg Army in northern Italy in 1734, Mercy and Württemberg commanded a force of approximately 50,000 men and brought 37,000 into battle at Parma in June.[7]

In the spring of 1734, the Austrian military conducted a military buildup in the region of Mantua, and charged Mercy, driving back French-Sardinian forces in the vicinity of Parma. Mercy, who struggled with failing health throughout the campaign, and Württemberg advanced from a position near San Benedetto in mid-May to threaten the allied army. After an advance of nine days, Württemberg crossed the Enza River, reaching the town of Sorbolo to the northeast of Parma. Mercy was stuck with illness at this period, and Habsburg emperor Charles VI dispatched Königsegg to help Württemberg with the management of the army. Both French King Louis XV and Charles VI hoped for a decisive outcome in this theatre and sent goading letters to their commanders. The result was a few days of intense skirmishing around Colorno in early June, with neither side gaining a decisive advantage.[8]

Parma: 29 June 1734

The decision came when Mercy, health somewhat restored, moved to the south of Parma, establishing a new camp to the southwest of the city on 28 June. Advancing from the region around Colorno, the French and their Piedmontese allies moved

---

    Prelude to the Study of Napoleon (Oxford: Claredon, 1927). For the best study in Italian, see the relevant sections of Ciro Paoletti, *Capitani di Casa Savoia* (Rome: USSME, 2007), pp.297–354.

6   Gaston Bodart, *Militär-historisches Kriegs-Lexikon 1618–1905* (Vienna: Stern, 1908), p.180.
7   Bodart, *Militär-historisches Kriegs-Lexikon*, p.180.
8   John L. Sutton, *The King's Honor and the King's Cardinal* (Lexington: University of Kentucky Press, 1980), pp.165–166.

south to attack the Imperial Army on 29 June. The Imperial forces, likewise, moved north to attack the allies on the morning of the 29th. To the immediate west of Parma, near a small farming village called Crocetta, the largest battle in Europe since Denain, over 20 years earlier, was about the be fought.[9]

The result was a meeting engagement or encounter battle, where both the French and the Austrians fed forces into the fight on a relatively narrow front, bound by the glacis and bastions of Parma to the east. The landscape was typical of a rural farming region of central northern Italy. Agricultural fields, cordoned by hedges, expanding out from stone farmhouses and farmyards, or *cascina*. Charles Emmanuel de Warnery, a young officer in Sardinian service at this battle who would go on to fight for the Prussians and become one of the most prolific military authors of the eighteenth century, described the scene:

> The ground was very flat, with no elevation at the road ... everything was cut by ditches and crossed by rows of trees which supported hedges, as is quite common in Lombardy, the space between these lines is cultivated, and the grain was very high. In this country, you cannot see one hundred paces in front of you, because the vines are drawn from one tree to another like high hedges, and beside that the corn and hemp was over seven feet high.[10]

Often, we imagine the battlefields of the eighteenth century as flat, wide-open terrain. Even in relatively open agricultural land, this terrain is rarely wide open, as Warnery indicates. It was in this type of environment that the two armies attempted to come to grips with one another. Perhaps unsurprisingly, both sides made efforts at reconnaissance, in order to protect themselves from surprise in an area where visibility was often down to 100 paces. Along the roads, you could see farther. A French officer wrote after the battle, 'while we were examining the ground we saw at 8 o'clock a small troop of enemy cavalry which was stationed 500 paces further on, down the road to Piacenza.'[11] As a result, a cavalry vanguard was sent forward, ahead of the main vanguard of 36 grenadier companies, in order to 'reconnoitre, and the generals went with them.'[12]

The French generals, realizing a battle was imminent, immediately took steps to seize key pieces of terrain. A French *lieutenant général*, the marquis de Louvigny, was ordered to occupy the three *Cascina* farmyards of the village of Crocetta, and additional troops occupied the ditches alongside the lanes which ran east to Piacenza and northwest to Cremona, forming an angle which refused the French right flank. The French were now posted in a strong position, with relatively linear

---

9 Sutton, *The King's Honor and the King's Cardinal*, pp.168–169
10 Charles de Warnery, *Anecdotes et pensées historiques et militaires, écrites vers l'année 1774* (Halle: Jean Jaques Court, 1781), p.16.
11 The National Archives of the United Kingdom (TNAUK): SP 92/37, Relation de L'Affair suivie entre L'armee Imperiale et celle des Allies Le 29. June 1734 devant la Ville de Parme, f.248.
12 TNAUK: SP 92/37, f.248.

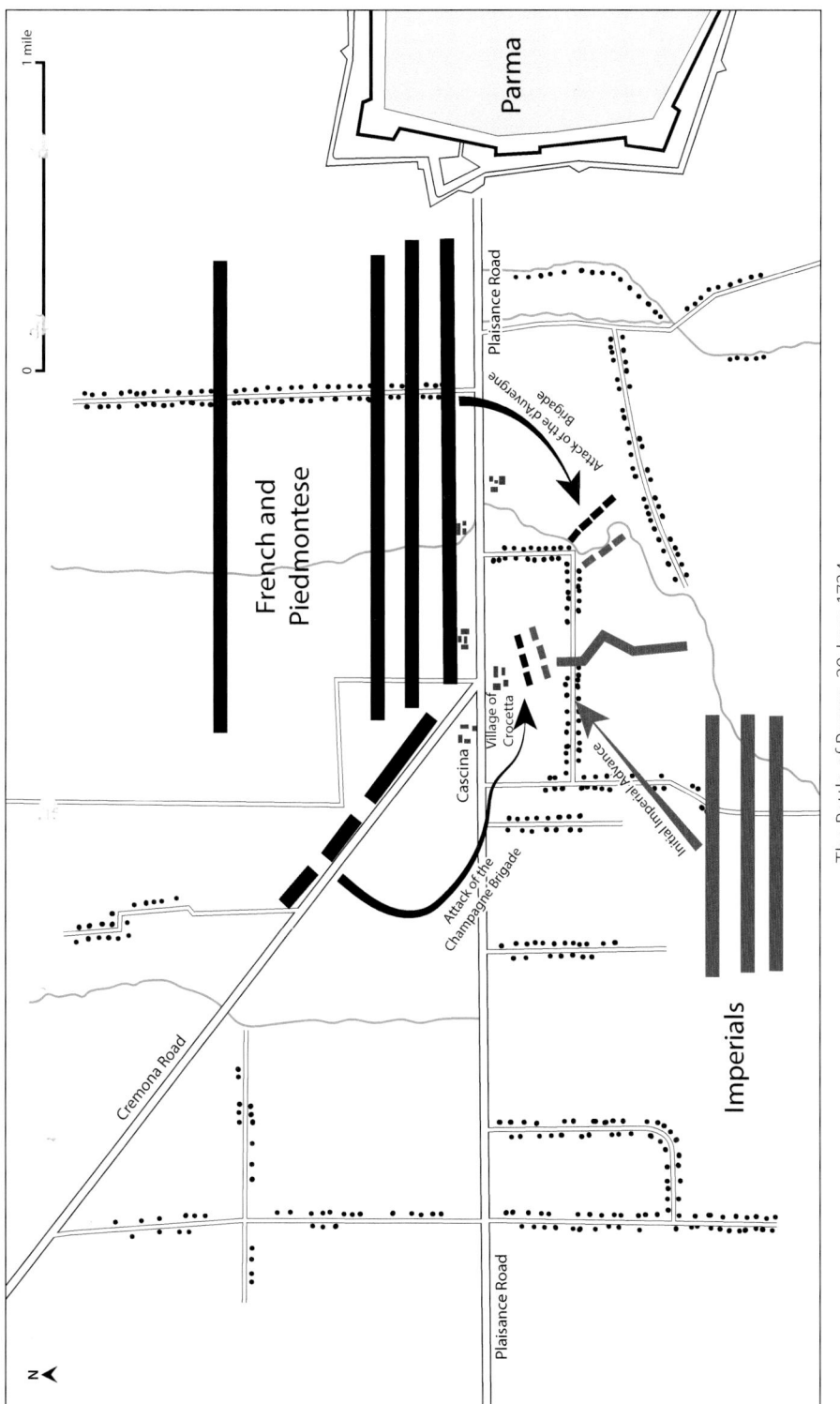

The Battle of Parma, 29 June 1734.

cover protecting the vast majority of their forces, and troops strongly fortified in stone strongpoints in the centre of their line.[13]

Not content to hold this position, the French also deployed skirmishers, or platoons of grenadiers acting as skirmishers, in front of their position. In all there were about 200 of these skirmishers, distributed in platoons, taking cover behind hedges, approximately 300 paces forward of the main French position.[14] Warnery notes a similar practice, at least as far as keeping contact with the enemy, generally, because of the difficult terrain: 'both sides detached runners and drummers from each troop of infantry, to observe the enemy closely without engaging, and to withdraw to their respective bodies on the enemy approach.'[15]

The Imperial army, advancing from the southwest, began to get some idea of this strong position. Württemberg tried to manage a reconnaissance with grenadiers and hussars while preventing the infantry in columns from piling up. He decided to pull some of his forces back, only leaving '100 hussars, 2 grenadier companies, and 400 men from the picket' in contact with the enemy.[16] Mercy was furious and this hesitation and the disagreement between the aggressive Mercy and more cautious Württemberg reached a fever pitch. Mercy declared: 'I will reconnoitre the enemy myself, at the head of my army!' commenting further to Württemberg, 'I see clearly enough and will do things my own way.'[17] At the very least, this was Württemberg's telling of events; Mercy did not live long enough to give his side of the story, and Württemberg's own days were numbered.[18]

By now, the good positions of the French and Sardinians were obvious. The *Obrist* of the Althann Regiment reported back to the Austrian command: 'it seemed to him that the entire enemy army was posted in ditches and cascines directly to his front.'[19] Despite the mounting evidence of a large enemy force in his front, Mercy remained convinced it was only a small detachment. By this point, both Imperial commanders were talking past each other, if they responded to the other at all.[20] Württemberg continued to carefully deploy his forces for a full attack, ignoring Mercy's repeated jibes – 'you must attack, it is only a small corps' – before Mercy finally led the small group of assembled forces forward himself.[21]

Mercy's attack first ran into fire from the grenadiers who had advanced as skirmishers in the hedges around Crocetta. At the main French position, the first

---

13 Kriegsgeschichtlichen Abtheilung, *Feldzüge des Prinzen Eugen von Savoyen* (Vienna: Verlag des K.K. Generalstabes, 1891), vol.19, p.344.
14 Kriegsgeschichtlichen Abtheilung, *Feldzüge des Prinzen Eugen von Savoyen*, vol.19, p.345.
15 Charles de Warnery, *Anecdotes et pensées historiques et militaires, écrites vers l'année 1774* (Halle: Jean Jaques Court, 1781), p.16
16 Württemberg's report to Charles VI, quoted in Ludovico Oberziner, 'La Battaglia di Parma, 29 giugno 1734', in *Atti del Congresso internazionale si sxienze storiche* (Rome: Tipografia Della R. Accademia dei Lincei, 1906) p.451.
17 Württemberg's report to Charles VI, *Atti del Congresso internazionale si sxienze storiche*, pp.451–452.
18 Sutton presents this exchange as fact, which says something about Mercy's reputation. Sutton, *The King's Honor and the King's Cardinal*, pp.169.
19 Württemberg's report to Charles VI, *Atti del Congresso internazionale si sxienze storiche*, p.453.
20 Kriegsgeschichtlichen Abtheilung, *Feldzüge des Prinzen Eugen von Savoyen*, p.345.
21 Württemberg's report to Charles VI, *Atti del Congresso internazionale si sxienze storiche*, p.453.

indication that an attack had begun was the firing from the grenadiers. A French officer remembered, 'while we were in the midst of everything, the advanced platoons of grenadiers began to fire and retire.'[22] The Imperial troops brushed aside the skirmishing grenadiers, and advanced directly on the cascines, some of which were cleared. Württemberg recalled: 'The Brigade did their work with such steadfastness and bravery, and the enemy was immediately dislodged from 3 ditches and 3 cascines. But because they were unsupported… they were unable to drive the enemy, who poured a terrible fire from two large stone cascines about 40 to 50 paces away.'[23]

Mercy followed this attack closely, riding along a small sunken lane parallel to the Piacenza. This road was swept by the fire of both the French grenadiers in the cascines and the fire of the Picardie brigade. Mercy was promptly killed.[24] Command devolved onto Württemberg, who calmly continued to form his infantry for an advance.

Württemberg's first attack drew praise from his adversaries, who noted the cool way that the Imperial infantry behaved under fire. One French officer recalled the 'great order' of the Imperial advance.[25] Another, Félix François d'Espié, recalled that the Imperials advanced 'with such order that it looked like a peacetime exercise.'[26] Warnery recalled, 'The imperial army advanced with commendable order, fusils at the shoulder, and received two discharges from the enemy without responding.'[27]

A seesaw battle raged for the rest of the day, but the lines were essentially set. Occasionally they would move, during counterattacks like those launched by the Picardie and Champagne brigades, but the rest of the day was essentially an infantry firefight, conducted at relatively close range, with both sides in a mix of buildings, ditches, hedges, and entrenchments – not in ranks, but out in the open. A French officer recalled, 'The armies were both close, taking cover behind ditches, with a narrow space in between, so they could shoot each other at close range.'[28] This phrase, 'so that they could shoot each other at close range', seems to imply that such a sustained, close-range firefight, was only facilitated by the type of terrain involved. Indeed, if we turn to sources which describe the quality of the firefight at Parma, it does indeed appear that contemporaries found it odd.

The Infantry Firefight at Parma

The first aspect of the Battle of Parma that contemporaries felt the need to comment on was the lack of cavalry involvement. This was a real fight between opposing

---

22 TNAUK: SP 92/37, f.249.
23 Württemberg's report to Charles VI, *Atti del Congresso internazionale si sxienze* storiche, pp.453–454.
24 Kriegsgeschichtlichen Abtheilung, *Feldzüge des Prinzen Eugen von Savoyen*, vol.19, p.347.
25 TNAUK: SP 92/37, f. 249. *Relation de L'Affaire suicie entre L'Armee Imperiale et celle des Allies, Le 29. Juin 1734.*
26 Félix François d' Espié, *Memoirs de la guerre d'Italie* (Paris: Unknown, 1777), p.166.
27 Charles de Warnery, *Des Herrn Generalmajor von Warnery Sämtliche Schriften* (Hannover: Helwingischen Verlag, 1785), vol.2, p.204.
28 Société Savoisenne D'Historie et D'Archéologie, *Mémoires et Documents* (Chambéry: Imprimerie Ménard, 1896), vol.35, p.xv.

groups of infantry, and infantry only. One French officer recalled that the in this type of terrain, horsemen 'were afraid of finding infantry there, which cause them to withdraw.'²⁹ Another report commented, 'it is to be observed the infantry alone on both sides has suffered, while the horse were only spectators.'³⁰ Another, more direct French soldier commented on the horse at Parma: 'The cavalry did nothing, it was not a country for them.'³¹ Artillery remained involved, however, providing supporting fire, particularly against troops sheltering in the cascina.

Contemporaries also felt the need to explain that almost no bayonet fighting occurred at Parma. Warney recalled, 'the whole battle was fought with small arms fire: the Germans [Imperials] once advanced with cold steel… but were driven back by a discharge… the French showed not the slightest desire to charge with lowered bayonets.'³² In different work, Warnery recalled the heroic actions of an officer named Benoit Lucadou, who killed an Imperial grenadier officer in hand-to-hand combat. That officer crossed one of the many ditches covering the battlefield, trying to motivate his reluctant grenadiers to engage in close combat.

> As the grenadiers were hesitant to pass the ditch, quite deep, one of the officers, no doubt a fire-eater, descended, and went back up in order to urge them to follow his example, with no other weapon but the sword in his hand… Lucadou, an intrepid man, although not five feet tall, approached him and giving him a bayonet thrust, at the same time shot him through the body and he fell. The fall of the German dismayed his soldiers so much that they no longer thought of crossing the ditch, not even to retrieve his body. I could cite other examples where an officer was killed, and his troops were completely dismayed.³³

This imperial grenadier officer earnestly desired to lead his men into close combat with the enemy, as many eighteenth-century officers did. In doing so, he was willing to expose himself to personal risk, leading from the front. The result was his death, demoralizing his men, who continued in their hesitancy to engage in hand-to-hand combat. This is a perfect example of the concept of negotiated authority which lay at the heart of eighteenth-century infantry combat. The officer wanted his men to do something. They initially refused, so he took action to motivate them to follow orders, and in the end, failed to set them in motion. These soldiers, like many of their comrades on the battlefield at Parma, were hesitant to engage in melee combat, as many other observers noted.

A collection of contemporary Italian reports on the news of 1734 reported:

---

29 TNAUK: SP 92/37, f.250.
30 TNAUK: SP 92/37, f.252, Arthur Villettes to Newcastle, 3 July 1734.
31 Société Savoisenne, *Mémoires et Documents*, vol.35, pp.xv–xvi.
32 Charles de Warnery, *Sämtliche Schriften*, vol.2, p.204.
33 Charles de Warnery, *Anecdotes et pensées historiques et militaires, écrites vers l'année 1774* (Halle: Jean Jaques Court, 1781), p.17.

...the infantry troops themselves had no field of fighting for the sword in hand, or with the bayonet at the mouth of the fusil, due to the fact that they were separated by a canal, or a very deep ditch. But the musketry fire was continuous and fierce, unlikely that soon as there will be an example of another like it in the histories.[34]

An anonymous French officer at the battle recalled: 'the fire was very lively without there being hand-to-hand combat.'[35]

Contemporaries were surprised by the lack of hand-to-hand combat because melee and the threat of melee had been so decisive in their immediate military history. Most battles of the Nine Years War and War of Spanish Succession included some element of melee combat, particularly when fighting at close quarters for possession of villages. Likewise, while infantry often bore the brunt of the fighting, cavalry proved itself as an important weapon of decision in the Duke of Marlborough's battles. Likewise, in 1734, the combatants were barely 15 years removed from the death of the Charles XII of Sweden, whose Caroliner and their *Gå-På* (head-on) tactics had stormed Russian, Danish, Saxon, and Polish positions in the Great Northern War. The Chevalier Folard had only just published his commentaries on Polybius, suggesting the utility of columns in warfare. Though it might seem obvious from the vantage point of the middle of the twenty-first century that firearms would come to fully dominate the battlefield, it was not yet apparent to the combatants at Parma or the community of military Europe reacting to the news of the battle. Cavalry and the bayonet had proven effective time and again in the very recent past, and contemporaries were surprised by their lack of use at Parma.

If firearms were the primary method of combat at Parma, how did contemporaries use them? What tactics were employed by each side in the battle? Although the French-Piedmontese and the Imperial armies began the day using very different tactics, by the end of the day, both had switched to the French method. The methods initially used by the Imperials will be more familiar to those with a conception of the 'standard' model of an eighteenth-century firefight, so we will begin there.

Charles Emmanuel de Warnery, observing the 'commendable order' of the Imperial advance, described their tactics: 'Arriving one hundred paces from the enemy, instead of continuing to advance, they halted, and began to a regular and sharp fire.'[36] Writing with benefit of hindsight in the 1770s, Warnery noted the slowness of the Imperial fire, which he attributed to the continued use of wooden rammers, and then argued that the Imperial commander, 'realizing that the enemy return fire was weak, in the manner of Croats, [the Imperial infantry] marched forward with the beating of drums. The soldiers, wearing their jackets, presented a fine sight.'[37] Warnery confirms that early in the day, during the initial attack, the Imperials were fighting in a very conventional manner.

34　Anon., *La storia dell'anno 1734, Divisa in quattro libri* (Amsterdam: Unknown, 1735), p.24.
35　TNAUK: SP 92/37/250.
36　Warnery, *Anecdotes et pensées historiques et militaires*, p.17.
37　Warnery, *Anecdotes et pensées historiques et militaires*, p.17.

Another officer with the Franco-Piedmontese Army likewise recalled that the Imperials: '...formed on a considerable front, the first rank fired, they then returned to the rear, others advancing and firing on the same ground. They perpetually continued this maneuver with such order that it looked like a peacetime exercise. Despite the sound of our musketry, and theirs, we could distinctly hear their officers shouting, "Feuer!"'[38]

Although it is not exactly clear what is transpiring, the Imperials seem to be engaged in a sort of fire-by-ranks or street-firing (*Gassen-Feuer*), where the first rank gives fire, retires, and then their places are taken up by the next rank.[39] The image of troops doing this perfectly, as if on a peacetime exercise, matches well with Christopher Duffy's repeated assertion (maintained in this work) that the only troops who would usually maintain peacetime regularity in combat were new soldiers after a long period of peace.[40] This strict order broke down in the course of the day.

The French, from the outset, acknowledged that maintenance of peacetime order was unlikely. Instead, in defending their position, they decided to allow their men to fire independently. A French officer described:

> Our firing was not as methodical as theirs but was not less deadly; we had given our men the freedom to fire at their pleasure, after they had used their cartridges, they scooped powder by the handful, loading it, and the ball, then primed, knocked the butt on the ground, and fired. Such at least was the method of the five battalions of our brigade, sometimes the fire was slower than the enemies, sometimes quicker, but it was continual.[41]

It is possible that not all French brigades fought in this way. Warnery speaks of the enemy receiving two 'discharges' which implies some sort of volley fire.[42] Despite Warnery's comment, it appears that the vast majority of the French and Piedmontese fired at will. The theorist Chabot, writing in 1756, asserts that the French fired at will during the battle at Parma.[43] Both the French and the Piedmontese would keep this tradition of intentional independent fire alive later in the century.[44] The fight at Parma became a by-word for troops exchanging fire at independently, at close range, behind cover. Reporting from the field of battle at Carillon in July 1758, Louis-Joseph de Montcalm stated: 'At one o'clock the enemy vigorously attacked us in four columns, mixed up with their irregular troops and best sharp shooters. The fire on the one side and on the other was like that at the battle of Parma, and the fight continued until eight o'clock at night.'[45]

---

38 D'Espié, *Memoirs de la guerre d'Italie*, pp.166–167.
39 For an explanation of *Gassen-Feuer*, see Duffy, *Instrument of War*, p.445.
40 Duffy, *Instrument of War*, p.446. See also, Duffy, *Military Experience in the Age of Reason*, p.214.
41 D'Espié, *Memoirs de la guerre d'Italie*, pp.166–167.
42 Warnery, *Anecdotes et pensées historiques et militaires*, p.17.
43 Le Comte de Chabot, *Réflexions critiques sur les differens systêmes de tactique de Folard* (Berlin: Neaulme, 1756), p.6.
44 Cerino Badone, *You Have to Die in Piedmont!*, p.104.
45 Brodhead and O'Callaghan, *Documents Relative to the Colonial history of the State of New York* (Albany: Wekd, Parsons, and Co., 1858) p.733. For repeated references to troops

This type of firefight was made possible by the extensive use of cover by the infantry on both sides. As Imperial troops advanced on their position, Warnery remembered that the French troops 'held on, keeping low to the ground, occasionally firing a shot.'[46] Another French officer recalled, 'All the rest of the day until 6 o'clock in the evening passed in continual fire on both sides, each of the parties being placed behind the ditches and the hedges.'[47] That at least part of the Imperial Army also engaged in this type of firing later in the day is supported by the assertion: 'The centre, with the infantry regiments… together 24 battalions in two lines, formed on the church of Valera against the canal via cava. A portion of the Grenadiers stood forward of this line and directed an unbroken and hearty fire of skirmishers towards the French.'[48]

What did this skirmishing fire look like in practice? Félix François d' Espié describes it as follows:

> [*Lieutenant Général*] de Louvigny made with his grenadiers a very lively and violent fire. Small groups of them would emerge from cover, as soon as they made their discharge, they were replaced by a similar number, again and again, those who had fired returned to load their weapons in the ditch, so that without being exposed, they made a continuous and ardent fire which taken together with the fire from the cascina Mambriani, was strong.[49]

Contemporaries thought that the fight at Parma was worth remembering, that it was a marked difference from previous engagements earlier in the eighteenth century as a result of the type of firefight that the infantry troops engaged in. On the French-Piedmontese side, this infantry fire was largely what later commentators would call a *feu de billebaude*, or independent fire.[50] On the Imperial side, an early effort to fire by complicated organizations of volleys or *Gassen-Feuer* fell by the wayside into a 'skirmishing' fire as the day wore on. This problem of maintaining complicated firings in actual combat would plague eighteenth-century armies for the entire period covered by this book.

Citing a lack of ammunition to continue the fight, Württemberg withdrew his army from the battlefield after dark on 29 June.[51] Both armies had suffered casualties: the Imperials just under 6,000 killed, wounded, captured and missing; the allies just over 4,000.[52] A detailed return of Austrian infantry and artillery casualties lists

---

alternatively firing at taking cover at that battle, see: <https://www.fortticonderoga.org/wp-content/uploads/2022/05/Preservation-and-Planning-Assessment-of-the-Carillon-Battlefield-Fort-Ticonderoga-PUBLIC.pdf>.
46 Warnery, *Anecdotes et pensées historiques et militaires*, p.17.
47 TNAUK: SP 92/37, f.251
48 Kriegsgeschichtlichen Abtheilung, *Feldzüge des Prinzen Eugen von Savoyen*, vol.19, p.349.
49 D'Espié, *Memoirs de la guerre d'Italie*, p.167.
50 Jacques-Antoine-Hippolyte, comte de Guibert, and Jonathan Abel (trans.), *Guibert's General Essay on Tactics* (Boston: Brill, 2022), p.78.
51 Württemberg's report to Charles VI, *Atti del Congresso internazionale si sxienze* storiche, p.454.
52 Sutton, *The King's Honor and the King's Cardinal*, p.169.

5,755 total losses: 2,085 dead and 3,670 wounded.[53] The Austrian-led force returned to their camp at San Benedetto near Mantua, the French enjoyed the fruits of their victory with a leisurely advance on Modena by way of Reggio; a small French garrison moved to occupy the surrendered city on 19 July.[54] Political manoeuvring and the vagaries of coalition warfare between Charles Emmanuel III and the French sapped their offensive of much potential.

The Road to Guastalla

The Austrians received a new commander on 11 July in the form of *Feldmarschall* Königsegg.[55] In a grand military tradition, Königsegg immediately began to advocate for more resources for his command. Although he wrote to Prince Eugene asking for money, his main complaint to Emperor Charles VI was not more money or men, but good officers:

> I should not spare your Imperial Majesty, how lacking we are in officers for Regiments of foot and horse. Some of my commanders have plenty of bravery and good will, but they are totally lacking in experience. It is even more so with the generals, many of whom were only Lieutenant-Colonels during the last war. Many have not served as staff officers, and so they don't know quite how to function in higher command roles. So, when we find ourselves in action in dense terrain, when a commanding general issues the first disposition, but cannot be everywhere at once in an attack through country divided by embankments and ditches, every officer needs to be able to think on his feet for himself.[56]

In this remarkable passage, Königsegg outlines the problems of command for infantry, but also in the army broadly, that were hampering the effectiveness of the army on the ground in the terrain of northern Italy. Despite these potential challenges, by 20 August *Feldmarschall* Königsegg felt his recently defeated army was ready to begin offensive probes of the French positions near Modena and Guastalla.[57]

The French had fortified their positions along the Secchia River, and assuming that the Imperial army would be quiet after the defeat at Parma, started to send their cavalry further away for better forage. As a result, their positions had been stretched dangerously thin even though they outnumbered the Imperial forces. Königsegg, somewhat buoyed up by money from Prince Eugene and reinforcements, decided

---

53 TNAUK: SP 80/227, f.70 'Des morts et des blesses de L'infanterie Imperiale au combat donné près de Parme'.
54 Kriegsgeschichtlichen Abtheilung, *Feldzüge des Prinzen Eugen von Savoyen*, Volume 19, p.361.
55 Sutton, *The King's Honor and the King's Cardinal*, p.170.
56 Kriegsgeschichtlichen Abtheilung, *Feldzüge des Prinzen Eugen von Savoyen*, vol.19, p.366.
57 Sutton, *The King's Honor and the King's Cardinal*, p.171.

on an attack. The result was the Surprise at Quistello during the night of 14–15 September 1734.[58]

The attack caught the French completely unaware: one French source indicated that the French troops were 'in the grapes and otherwise diverted.'[59] Fording the Secchia, the Austrian columns stormed through the French camp nearing Broglie's headquarters.[60] Broglie fled with his sons *en chemise*, or in his nightshirt without his pants.[61] The disaster resulted in the loss of about 2,000 French, killed and prisoners, with the rest fleeing into the night before reforming on the other side of Quistello.[62] One French account tried to downplay the loss, saying, perhaps unconvincingly, 'we have lost nothing except the first battalion of Guards, and the first Battalion of Savoye which we believe are prisoners.'[63] Another account was more honest regarding the loss:

> The French, based on the confidence that the Germans would not do anything, guarded the banks of the Secchia very badly, they were surprised at this point, the enemy was in the neighbourhood the Dauphin Brigade before they were fully aware. You can easily see the confusion that arose, several brigades immediately fell back, abandoning their camps and equipment… M. de Broglie only escaped with what he was wearing. The enemy was at this point a little confused, our infantry finally getting into a bit of order, formed up in an advantageous place, having a canal in front of them after abandoning the village of Quistello. The companies of Grenadiers of la Marine and de Rebinder were unable to withdraw, and were made prisoners. Riders were dispatched to round up the dispersed troops, so that a good number had returned by the end of the 15th. A council of war was called on the night of the 15th, the feeling was that M. de Broglie wanted to counterattack, the other generals were opposed, it was finally resolved to withdraw towards Guastalla.[64]

The cost of the affair on the part of Königsegg had only been about 400 men. He had inflicted 600 enemy casualties and made 1,500 prisoners, of whom 500 were officers. Moving north to San Benedetto, he followed the retreating French as far as Luzzara, where, on the 18th he decided to follow up on his success by attacking the French and Sardinian position at Guastalla.[65]

58　Kriegsgeschichtlichen Abtheilung, *Feldzüge des Prinzen Eugen von Savoyen*, vol.19, p.369.
59　TNAUK: SP 92/37, f.332, Extrait d'une lettre ecritte de Guastalla, le 17 Septembre 1734.
60　Kriegsgeschichtlichen Abtheilung, *Feldzüge des Prinzen Eugen von Savoyen*, vol.19, p.371.
61　TNAUK: SP 92/37, f.332, Extrait d'une lettre ecritte de Guastalla, le 17 Septembre 1734.
62　TNAUK: SP 92/37, f.333, Extrait d'une lettre ecritte de Guastalla, le 17 Septembre 1734.
63　TNAUK: SP 92/37, f.332, Extrait d'une lettre ecritte de Guastalla, le 17 Septembre 1734.
64　TNAUK: SP92/37, f.333–333a, Extrait d'une lettre ecrite de Guastalla, le 17. Septembre 1734.
65　Kriegsgeschichtlichen Abtheilung, *Feldzüge des Prinzen Eugen von Savoyen*, vol.19, pp.379–384.

Guastalla: 19 September 1734

Königsegg approached Guastalla on 19 September under the misapprehension that the allied army was already retiring across the Po over the bridge constructed next to Guastalla. Observing movement from a belltower in Luzzara, it seemed to him that the French were in retreat.[66] He believed that only a portion of the enemy force remained to his front, in a way similar to the false belief that Mercy had been under during the initial attack at Parma.[67] The generals were reluctant to shake the army into full array for battle because they thought that only 5,000 French remained on their side of the Po.[68] In actuality, only the baggage train had been sent over the Po, and an Allied confederate force of 40,000 Franco-Piedmontese awaited Königsegg in a position more formidable than that of Parma. Königsegg had approximately 30,000 men available for action as he moved toward Guastalla on the morning of the 19th.[69]

The allies awaited Königsegg along two causeways, sometimes called dams in the sources, which provided natural cover for the waiting allied infantry. The Austrian General Staff history of the war gives a detailed description of the terrain, closely matched by contemporary sources:

> The main road from Guastalla to Luzzara and direction of Imperial approach, follows the *argini maestro* 1200–2500 paces from the Po, past the villages of S. Giorgio [and] Tagliata… the space between the *argini maestro* and the Po, an elongated triangle tipped by Guastalla, is cut by a second, smaller dam, the *arginello,* on a path which also runs from Guastalla to Luzzara. Between these two [large] causeways lie several smaller ones, providing a defender with much natural cover or breastworks. To the west past the *arginello,* a narrow meadow stretches out to about halfway to Luzzara with a few paths, which allowed for the use of cavalry despite existing bushes and clusters of trees. North of this meadow was the forest of Luzzara. To the east of the *argini maestro* the terrain was vineyards and forests. Near the town and between the two causeways there were a few scattered villages, containing stone Cascines and a Capuchin monastery.[70]

Despite the setback at Quistello, the French felt confident in their defensive position. One letter from 17 September, after describing the disaster at Quistello, ends on a high note: 'the enemy is on their way here and I believe we are getting ready to receive them well.'[71] These causeways, the *arginello* and *argini maestro* or *argine*

---

66 Robert von Lindenbüchel, *Geschichte des Kaiserlich and Könglichen Infanterie-Regiments No. 35* (Vienna: F. Tempsky Verlag, 1897), vol.2, p.48.
67 Kriegsgeschichtlichen Abtheilung, *Feldzüge des Prinzen Eugen von Savoyen*, vol.19, p.384.
68 Kriegsgeschichtlichen Abtheilung, *Feldzüge des Prinzen Eugen von Savoyen*, vol.19, p.384.
69 Gaston Bodart, *Militär-historisches Kriegs-Lexikon 1618–1905* (Vienna: Stern, 1908), p.182.
70 Kriegsgeschichtlichen Abtheilung, *Feldzüge des Prinzen Eugen von Savoyen*, vol.19, p.380. For confirmation on the terrain from an eyewitness, see: d'Espié, *Memoirs de la guerre d'Italie*, pp.224–228.
71 TNAUK SP/37/333a, Extrait d'une lettre ecrite de Guastalla, le 17 Septembre 1734.

NORTHERN ITALY, 1734 31

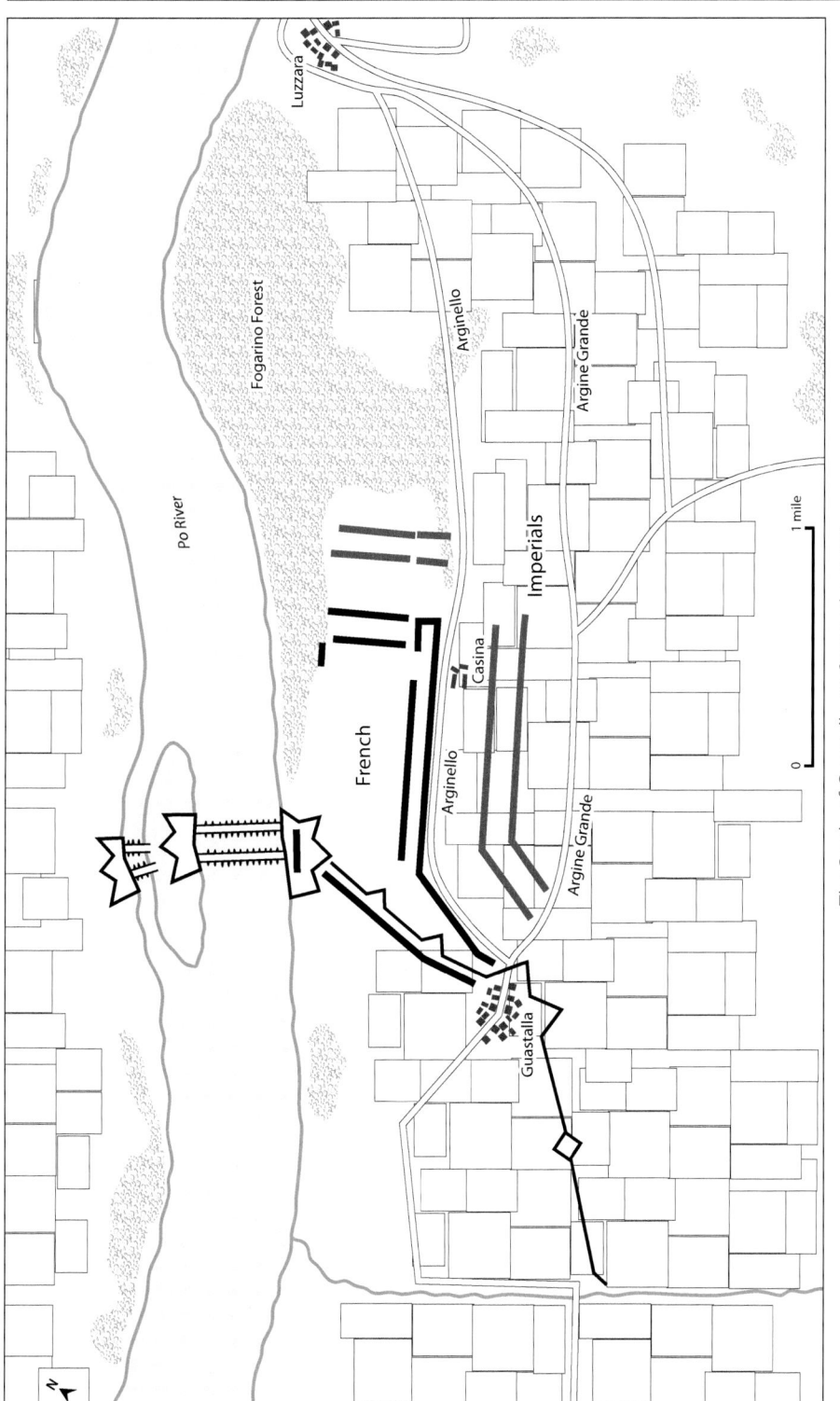

The Battle of Guastalla, 19 September 1734.

grande, were supplemented by fortified lines that the allies had constructed around Guastalla, in order to protect their bridge. A French veteran of the battle described the steps the allies took to further prepare this defensive position: 'In places where the hedges were not thick enough [for protection] our soldiers entrenched them with planks and pieces of wood taken from the nearby houses.'[72] At Parma, the allies had quickly seized advantageous ground before the battle. At Guastalla, they stood in an excellent natural defensive position that was supplemented by prepared positions.

Once again, allied grenadiers (Piedmontese in this case) were placed in the cascines of the surrounding villages in order to defend the position.[73] Contemporary reports praised Charles Emmanuel III for recognizing the strength of the position, saying:

> The King spent the night of the 16th and the day of the 17th on horseback calmly straightening things up. He learned on the 18th that the enemy, instead of stopping, wanted to follow their success. He posted the left of his army towards this river and the right towards Guastalla, making use of the advantage which the *argine* gave and resolved to wait for them. He was very happy on the evening of the 18th, when he saw him the situation that he was only able to be attacked with great disadvantage from to enemies.[74]

Königsegg began his attack at 10:00 a.m. on 19 September, coming from the north out of Luzzara, and led with a push by 12 grenadier companies towards the occupied cascines between the causeways, and an advance by the cavalry to the right (east) of the *arginello*.[75] The French cavalry, led by Alexis-Madeleine-Rosalie de Bois-Rogue, duc de Châtillon, drove the Imperial cavalry back after a sharp fight. Letters from French camp state that in a personal conference, Charles Emmanuel III made it clear to Châtillon that his defeat would lead to the collapse of the allied position.[76] After the initial defeat of the Imperial cavalry, the cavalry of both sides glowered at each other for the rest of the day and were content with an artillery bombardment conducted at a distance.

The Imperial infantry, led by 29 companies of grenadiers, cleared the cascines between the causeways, but was unable to press forward against the main body of the enemy, and returned to the cover the cascines and smaller causeways between the two large causeways. Imperial artillery began to deploy on the *argini maestro*, firing at the allies sheltering behind the *arginello*. This space is considerable: from the *argini maestro* at Tagliata to the *arginello* is three-quarters of a mile. Close to Guastalla, it is a third of a mile or less. Instead of numerous distinct attacks, as at Parma, both Königsegg and Charles Emmanuel III contented themselves with feeding troops into the firefight in the area between the *arginello* and *argini maestro*.

---

72   D'Espié, *Memoirs de la guerre d'Italie*, pp.227.
73   Kriegsgeschichtlichen Abtheilung, *Feldzüge des Prinzen Eugen von Savoyen*, vol.19, p.381.
74   TNAUK: SP 92/37, f.336, Du Camp de Guastalla.
75   Kriegsgeschichtlichen Abtheilung, *Feldzüge des Prinzen Eugen von Savoyen*, vol.19, p.385.
76   TNAUK: SP 92/37, f.336, Du Camp de Guastalla.

The Infantry Firefight at Guastalla

Although the cavalry had briefly engaged in hand-to-hand combat early in the battle, and Württemberg would be shot and killed trying to advance with cavalry late in the day, contemporaries once again express surprise that there was not an infantry melee at Guastalla. An officer describing the battle in French recalled, 'their cavalry having been defeated, their infantry tried to force ours from along the *argine*, but it never came to cold steel.'[77] The same source hammers this point home again, saying, 'In five hours of combat, the enemy retired without ever coming to cold steel.'[78] The veteran Warnery, who fought in the Piedmontese army in this battle, confirms later in the century, despite the opportunities, 'not once did the French go at the enemy with fixed bayonets and determination, even though they took much less artillery into the field than we do today.'[79]

Instead, as at Parma, the main method of combat between infantry was the fire of small arms supported by artillery. Qualitatively, the battles were quite similar. The British minister in Turin, having read reports of the battle, credited the allies with a victory due to superior exploitation of terrain: 'the loss of the confederates, tho considerable, was much less, great part of their troops being covered by the causeway and having the advantage of firing from a superior spot of ground.'[80] Instead of closing to close range, both sides maintained a firefight, likely at a greater distance than at Parma, where at times the fighting was between 50 to 150 paces. Warnery comments: 'At Guastalla they fought with fire from quite a distance.'[81]

A newly promoted *major* of Irish extraction, the 24-year-old Johann Sigismund Maquire von Inniskillen gives the best account of the desperate Austrian efforted to maintain fire control at Guastalla. Maquire served alongside Austrian Grenadiers under Prince Joseph of Saxe-Hildburghausen. Maquire described the scene:

> We marched slowly, with bayonets fixed, and when we were in the presence of the enemy, the Prince ordered me to gallop from company to company and forbid on pain of death to fire without orders from him or me. We make a bold front, and march in good order, until we come within 150 paces of them, when a grenadier fires and the others follow his example. The Prince shouts, sabres them, and makes incredible efforts to keep them from firing and to get them to advance. Count Grune and I do the like to the utmost of our ability. Prince Louis (Württemberg) runs up, sabres them also, and while performing the duty of a grenadier as well as that of a general… towards 4 in the afternoon we retired to the village where we were posted in the morning. No company mustered more than 10 men. Those wretches, who had nothing of the grenadier but the busby, got themselves slaughtered miserably, whereas if they had followed the orders of the commander, who

---

77 TNAUK: SP 92/37, f.336, Du Camp de Guastalla.
78 TNAUK: SP 92/37, f.337, Du Camp de Guastalla.
79 Charles de Warnery, *Sämtliche Schriften*, vol.2, p.205.
80 TNAUK: SP 92/37, f.326, Essex to Newcastle, 22 September 1734.
81 Charles de Warnery, *Sämtliche Schriften*, vol.2, p.205.

is a second Charles XII of Sweden, we would have not lost the tenth part. Instead of giving one volley and falling on the enemy with the bayonet, we did not kill more than thirty of them and our fire was of little service.[82]

Here, Maquire expresses the typical sentiment of an eighteenth-century officer who desired a quick resolution to an engagement between enemy infantry. He criticized the honour of the troops who fired (in his view) too early, and vastly preferred to attack with the bayonet over a long and indecisive firefight.

Instead of reaching a decision with bayonets, the commanders fed more and more infantry troops into a firefight, and the race became which reinforcements would arrive first or take up an advantageous position. Here, Charles Emmanuel III's leadership was decisive, as he spent the entire day riding from crisis point to crisis point, attempting to rally his men and feed reinforcements into the battle. An Italian letter from the battlefield comments:

> His Majesty commanded the army, and he rode all the lines, where storms of bullets could be heard, enlivening now the French, now ours, the men were so enraptured by his presence and valour, that they did nothing but accompany him with constant shouts of *Vive le Roy*, and with loud protests that they wanted to use every drop of his blood for his service.[83]

A decade later, military theorists were still citing Charles Emmanuel III's presence at Guastalla as an example of a general motivating his troops. A French military manual published in Turin recalled: 'Nothing gives soldiers more courage than to see their general exposed to the same dangers as they are. This was last seen at the Battle of Guastalla, where the presence of the King restored two entire brigades which had taken a sudden fright, and by his valour so animated to the troops that they won a complete victory.'[84]

As much as it might flatter the House of Savoy to think that Charles Emmanuel III was solely responsible for this victory, a French veteran recalled a slightly different motivation on the part of the French: 'Our soldiers yielded nothing to them, they defended themselves with infinite stubbornness and courage, saying loudly that they wished to gain this battle, so they might win back from the enemy the haversacks they had lost at Secchia [Quistello].'[85] Whether the troops were motivated by royal genius or the desire to regain personal property, it is difficult to escape the fact that Charles Emmanuel III brought reinforcements into the battle effectively. A French letter from the battle credits Charles Emmanuel with noticing 'that the enemies did not extend their left beyond *argini*, so he reinforced…with the Dragoons and infantry of Piedmont, who occupied the *argini maestro* (*le grand argine*) and took

---

82 Historical Manuscripts Commission, *Report on Manuscripts in Various Collections* (Hereford: Hereford Time Limited, 1913), vol.8, p.408.
83 TNAUK: SP 92/37, f.326: Copia di lettera scritta dal campo di Guastalla li 19. Settembre 1734. Emphasis in original.
84 De Bubilian, *La Science de la Guerre* (Turin: De L'Imprimerie Royale, 1744), p.195
85 D'Espié, *Memoirs de la guerre d'Italie*, p.238.

the enemy in the flank and rear.'[86] As a result of effective reinforcement, Charles Emmanuel III had his enemies surrounded on three sides.

Despite not wanting to charge into hand-to-hand combat, at Guastalla both sides, whether on the attack or reinforcing their positions, also showed a great willingness to move at speed, or run, on the battlefield. A young French officer in the 1740s, Jacques de Mercoyrol de Beaulieu, who fought in the Picardie brigade, was regaled with tales of both this urgency and Charles Emmanuel's presence by the old officers of the regiment:

> [The] old officers of the regiment, who had fought in the Italian campaigns in 1733, 1734, 1735, [said] at one of these battles (which were all victories) the King of Sardinia commanded there personally, and the Picardie brigade at its post on the right, was summoned by this monarch to move up towards the centre, where the main fighting was. They went almost at as in a race, the quickest arrived first; this eagerness awakened courage and gave new strength to the tired fighters, and the musketry grew more lively...[87]

These old officers, evidently proud of their role in the fighting of 1734, were extolling the advantages of moving at speed on the battlefield over the introduction of cadenced marching. Imperial troops, having learned from their parade ground advance at Parma, also fought in a more realistic manner. A French artillery officer recalled that at Guastalla, a battery supporting the Champagne Regiment started to run low on round-shot ammunition and slackened its fire as a result:

> The enemy noticed this very quickly, and resolved to seize the battery, which had stopped their advance up to that point, and push back the troops which defended it. This advance was made in good order and at an almost precipitous pace: assured of success. On their approach one of the officers ran to the ammunition and seized some canister charges. These guns were promptly loaded with a sufficient quantity of these balls and fired at the Germans at close range to murderous effect; they instantly turned and fled.[88]

This account highlights a number of features: the Imperial infantry, hanging back (likely behind the cover of a minor causeway) notices the slackening fire of a battery, and decided to move up as a result. This movement is performed at an 'almost precipitous' speed as a result of the danger of remaining open in front of a battery. Despite this speed, the battery opens fire, and the infantry immediately flee the exposed position they find themselves in. This emphasis on cover is returned to again and again by the officers who saw the fight at Guastalla or spoke to older

---

86   TNAUK: SP 92/37, f.337, Du Camp de Guastalla.
87   Jacques de Mercoyrol de Beaulieu, *Campagnes de Jacques de Mercoyrol de Beaulieu, capitaine au régiment de Picardie: 1743–1763* (Paris: Libairie Renouard, 1915), p.184.
88   Dupuget, *Essai sur l'usage de l'altillerie, dans la guerre de campagne & dans celle de sieges* (Amsterdam: Arckstée and Merkus, 1771), pp.8–9.

officers who had. Warnery recalled, 'all four ranks laid down and fought in the manner of croats.'[89] Jacques-Antoine-Hippolyte, comte de Guibert, when writing his essay on tactics later in the century, cited Guastalla as evidence for his position that soldiers should never be allowed to kneel when firing: 'I do not allow the position of … kneeling on the ground… in addition, on the approach of the enemy, it is a posture that soldiers cannot often be made to quit. At Parma and at Guastalla, nearly all our infantry and that of the Austrians fought thus.'[90] The Imperial troops even took this desire for cover a step further, actually entrenching on the battlefield while the battle was in progress. The senior French officer on the field, Coigny, ended his report to the minister of war: 'I have not yet mentioned a remarkable occurrence, that even during the action, the Germans were working to entrench themselves.'[91]

At Guastalla, cavalry played a more important role than at Parma, with a field in the eastern portion of the battlefield being suitable for cavalry combat. Here, Württemberg was killed while trying to lead his cavalry to the attack late in the battle. Despite this, the most important part of the action was still a firefight, with the French bringing barrels of gunpowder to the fore to allow the troops to restore their supply of ammunition.[92] Despite Königsegg trying a number of ideas, including loading troops onto boats and rowing behind the allied front line, the Imperial troops could not break through the confederate position on the *arginello*, and by late in the day outflanked and running low on ammunition, they withdrew. The Imperials, having learned at their own expense, covered their retreat effectively with grenadiers placed in cascines to delay the allied pursuit.[93]

This was the last major field battle of the War of Polish Succession in Italy. The two armies drew apart, and Austrians managed to prevent the fall of the fortress of Mirandola later in the year.[94] Guastalla was also the last major battle in a confrontation between European powers excluding the Ottomans until the Battle of Mollwitz, more familiar to anglophone readers due to its association with Frederick II of Prussia. As a result of their place in a seemingly sluggish *Kabinettskriege* without the presence of a famous general to lead either army, the fighting in northern Italy in 1734 has been mostly ignored for the last hundred years of historiography, a few monographs by Italian authors notwithstanding.[95]

This is most unfortunate, as the infantry combat in northern Italy in 1734 displays important themes which stretch out over the next 50 years of European and Atlantic military history. During this period, infantry frequently engaged enemy at longer ranges, perhaps 100 to 200 yards away. Despite this distance, they tried to aim their shots. These men would fight as skirmishers when in certain armies and circumstances. Infantry frequently fired at will, or without orders, particularly after an initial volley. Infantrymen would move with speed on the battlefield, running or

---

89 Charles de Warnery, *Sämtliche Schriften*, vol.2, p.211.
90 Guibert and Abel (trans.), *Guibert's General Essay on Tactics*, p.74.
91 Pajol, *Les Guerres sous Louis XV* (Paris: Imprimeurs de L'Institut, 1881), vol.1, p.521.
92 D'Espié, *Memoirs de la guerre d'Italie*, p.238.
93 Kriegsgeschichtlichen Abtheilung, *Feldzüge des Prinzen Eugen von Savoyen*, vol.19, p.391.
94 Sutton, *The King's Honor and the King's Cardinal*, p.175.
95 See, for example the short monograph: Andrea Santangelo, *Guastalla 1734: Un battaglia per il trono di Polonia*, (Italy: Verba Martis, 2003).

jogging to rapidly close to distance to a target or get out of the line of fire. That same infantry would sometimes hug the ground, trying to minimize themselves as a target. Sometimes the men would fight from a prone or kneeling position. Finally, despite the image of bayonet fighting as incredibly common, eighteenth-century soldiers displayed an extreme hesitancy to engage the enemy in melee combat with bayonets when using firearms was a possibility. Officers frequently bewailed these developments in their writings, as they sought to create an ideal war machine which could function like clockwork. This tension, between the officers' demands and soldiers' actions, lies at the heart of the negotiated authority which actually drove the tactical developments of eighteenth-century infantry battle.

The fighting at Parma and Guastalla in 1734 shows all of these common themes of eighteenth-century infantry combat. In many ways, these battles inaugurated eighteenth-century infantry combat. On most eighteenth-century battlefields, infantry displayed portions of the flexibility shown here, particularly when the threat from enemy cavalry was negligible. However, another major development in tactics, this time from Northern Europe, prevented this style of fighting from becoming the norm. As a result, many officers believed that the best way to manage infantry combat was to bring the men more tightly under control. As a result of this belief, the flexible practices and methods of Parma and Guastalla were overshadowed by military developments just six years later, when Frederick II of Prussia led a tightly drilled military to invade the Austrian province of Silesia. The Prussian adventure in Silesia, and the fame that it brought to both its monarch and its army, caused Prussian military practices to dominate the theoretical world of infantry battle between 1740 and 1790. The flexible infantry tactics shown at Parma and Guastalla would be utilized (even by the Prussians), but another more rigid ideal would overawe writing about eighteenth-century warfare from that time to the present.

2

Armies, the Experience of Battle, and Tactics

The battlefields at Parma and Guastalla contain many interesting vignettes of eighteenth-century infantry combat, but how did these armies actually function? Explaining this process is the task of the next three chapters, which will examine the structure and tactics of these armies, the background of the men who composed them, and their officers' desire for rigid tactical control. This chapter demonstrates that in terms of structure and tactics, there were broad similarities across many eighteenth-century armies. Each of these armies possessed unique characteristics but fought in a way that was mutually intelligible across military Europe as a whole. Soldiers might be terrified by the discharge of an enemy artillery battery, overawed by the battalion volleys of enemy infantry, or shaken by the thundering hooves of enemy cavalry, but they rarely experienced a style of war that they did not comprehend as part of their own military experience. As a result, officers fiercely debated micro-tactical decisions: was it best to reserve your battalions fire and intimidate the enemy? Should you open fire at longer range with platoon fire? How many platoons should compose a battalion? These debates occurred across military Europe, and across the eighteenth century, as officers grappled with the tactical demands of linear warfare. As their officers debated best practices, the soldiers who composed the vast majority of these armies drilled, worked, fought, and died. They had their own ideas about the best way to fight and would not always wait for officers to tell them what to do.

Army Life and the Soldier

Army life was difficult for these enlisted men. They would march an average of 14 miles a day, in straight last shoes that wore out easily.[1] In the Itzenplitz Regiment of the Prussian Army, a soldier reckoned, that they 'marched more than two hundred miles from autumn to now, and every 3rd soldier is without stocking or shoes.'[2]

---

1   This average was arrived at by examining commentaries by authors such as Clausewitz and Jomini, primary sources cited in the works of Christopher Duffy (the majority come from *By Force of Arms*), *The Journal and Order Book of Captain Robert Kirkwood*, *The Papers of George Washington*, in addition to manuscript sources. Straight last shoes were leather soled and not distinguished into right and left. They were hard on the feet.
2   Christian Zander, *Fundstücke: Dokumente und Briefe einer preussischen Bauernfamilie* (Hamburg: Kovacĺ, 2015), p.68.

Their caloric intake was high when compared with civilian life, and contained more servings of protein.³ This higher food intake was matched by the intensity of their army labour. Usually, the day would begin quite early. The Prussian regulations indicate that Reveille was beaten when visibility reached 40 paces. In some armies, soldiers would immediately fall in for an accountability formation, to make sure that no one had deserted during the night. After the role was called on a campaign, the mess group would immediately begin to perform its daily tasks. In a peacetime garrison, the whole body of men might perform drill after cleaning their uniforms and equipment, depending on the day of the month.⁴

Throughout the eighteenth century, soldiers dressed in broadly similar ways. Starting with the feet, woollen stockings or socks would be pulled up over the knees. Soldiers in some armies wore footwraps. The long linen shirt was both shirt and undergarment for these men. Over their shirts and stockings, soldiers wore woollen or linen breeches that buttoned at the waist and were buttoned or tied at the knees. The lower legwear would be completed, increasingly after 1700, with long gaiters which buttoned from the feet to above the knee. American, British, and German-speaking subsidy troops in North America during the War of Independence did away with breeches and gaiters and wore a long skinny trouser which buttoned at the ankle, called 'gaitered trowsers' or 'overalls.' Like the breeches and gaiters this garment replaced, it could be made of linen in summer and wool in winter.

Wearing his legwear and shirt, a soldier would be ready for most of the summer's hard daily work of army life, described below. In colder seasons, on the drill square, on the march and in battle, he would also wear his waistcoat and regimental coat. The waistcoat was a woollen vest, to which armies that fought in colder climates attached sleeves for exclusive use in summer, discarding their regimental coats. The regimental coat was a more elaborate affair, with lapels, cuffs, and a collar. In many armies, the lapels or facings were functional and could be buttoned over for greater warmth in winter. Around his neck, the soldier would wear a stock or neck cloth, usually less onerous than is claimed, often in black or red, which would hide the collar of the shirt. The cocked hat, or three-cornered hat, would complete the ensemble, worn over the soldier's own long hair, usually not a wig. Specialist troops, such as grenadiers or fusiliers, would often wear a metal or fur fronted cap, designed to exaggerate the soldier's height.

Uniforms were a collection of blues, reds, whites, and greens. Why would this be the case? Generals favoured highly visible and identifiable uniforms because they allowed troops to be recognized, controlled, and moved. Units wore brightly coloured coats, and different coloured lapels and turnbacks (coat tails or skirts), that allowed officers and men to distinguish between different units of the same army. Soldiers and officers effectively utilized their clothing and equipment in order to fight as efficiently as possible. Ironically enough, it is only after the Seven Years War that some European armies became so infatuated with their perception of the external trappings of the Prussian Army. Thus, in the late eighteenth century, some

---

3   Fred Anderson, *A People's Army: Massachusetts Soldiers and Society in the Seven Years' War* (Chapel Hill: University of North Carolina Press, 1984), p.84.
4   Duffy, *The Army of Frederick the Great*, p.80.

officers argued for formality without function. They would have been rather out of place in the Europe of 1757, or the North America of 1777. Now, we might also ask, were their brightly coloured uniforms restrictive?

Clothing in various eighteenth-century militaries was undoubtedly more restrictive of movement than military clothing after the mid-nineteenth century. With that said, the clothing of eighteenth-century soldiers did not greatly hamper their efficiency in combat. Though still an intensely physical experience, eighteenth-century combat was on average less physically demanding than combat today. In an example of this logic, Christopher Duffy asserts that loads in the eighteenth-century averaged about 60 pounds, while modern soldiers in Iraq and Afghanistan carried around 120-plus pounds. Despite this, eighteenth-century warfare could still be incredibly physically demanding, as the 12-mile run of the Grenadiers of the 45th Regiment from Philadelphia to Germantown shows us. Likewise, Prince Henry and his army marched almost 100 miles during three days in August of 1760.

Officers and soldiers were concerned with the functionality and durability of the garments fighting men wore. After the Seven Years War, the Prussian Army completed the transition to woollen gaiters as a result of their functionality. *Generalleutnant* Schmettau reported:

> The Gaiters: They had formerly been made out of twill, but it is better that they be made out of cloth, experience has taught us that these are warmer and lay better, so that the soldier looks more orderly, therefore, they have been almost universally adopted. Although they cost twice as much as the others, they only need to be issued once a year, because they are much more durable than the others, and the company proprietors prefer to issue something that looks better.[5]

Finally, it is indisputable that soldiers took an interest in their uniforms, even the minor details. Period treatises, such as Cuthbertson, make it clear that officers cared a great deal about the uniforms of their men. Uniform details often became wrapped up in matters of honour, and as a result, ordinary soldiers came to focus on the symbology of their clothing. In 1787, when the second battalion of the 42nd (Royal Highland) Regiment was to be designated the 73rd (Highland) Regiment, the men complained that they would lose their royal facings (a deep blue colour.) Norman Macleod reported:

> I embrace this opportunity of sending you a Return of it, and of giving you a full account of its present state... I shall now speak of the clothing. As the Reg't we had the honour to have Royal Facings from the beginning and have done nothing to forfeit that honour, but on the contrary has been distinguished by brave behaviour, and severe sufferings, it hopes that tho separated from the Fourty Second, it will still be a Royal Highland Regt. It is not easy for me to express the anxiety felt on this account by the whole

---

5   Fredrich Wilhelm von Schemttau, *Einrichtung des Krieges-Wesens für die Preussische Infanterie zu Friedens-Zetien* (Berlin: Duncker and Humboldt, 2019), p.209.

corps. The officers certainly felt it as a point of honour, and on a mischievous report being raised that the facings were to be changed, the men loudly expressed their grief and rage. I must therefore earnestly recommend this point to your most serious consideration.[6]

Unfortunately for the troops, the petition was unsuccessful. We have now dressed our soldier, and we must arm him. Over his typical uniform the soldier would be crisscrossed by a number of belts. His sword belt or bayonet carriage would be worn above the waist or over the right shoulder. His ammunition or cartridge pouch would be worn over the left shoulder, with the pouch resting on the right hip for ease of access by the right hand. Over these, particularly on the march, the soldier would carry a linen haversack for food storage, a hair-on cowskin bag for his personal affects, and in some armies, a canteen. Armies varied between a larger canteen for every six men, and a small individual canteen for each soldier. Soldiers who did not carry the large canteen would be saddled with other mess equipment, such as a kettle, axe, or pick. In some armies, each soldier might also carry a few tent stakes.

Soldiers would often wear a small sword or hanger, more for decorative than combat purposes. The honour of the unit was associated with these short swords. If the troops performed poorly in battle, swords could be taken from them, as the Prussian Anhalt-Bernberg Regiment found out in the campaign of 1760. All soldiers would carry a bayonet, which could be locked to the end of their musket. In many armies, this attachment was permanently fixed. The standard infantry weapon was a smoothbore musket with a flintlock ignition system, usually of about .69 inches in calibre. For a weapon system notorious in a modern context for its inaccuracy, it would hit a man-sized target at 100 yards with relative accuracy and precision.[7] Furthermore, in battle, as opposed to the war of posts, the target would rarely be an individual, but rather a mass of infantry or cavalry over 100 yards in width. The weapon was loaded as follows. Our heavily laden musketeer would grasp a paper cartridge with gunpowder and ball from his pouch with his right hand and open it with his teeth. He would then prime the pan of his musket lock, a small trough enclosed to contain powder. He would then shut the frizzen (called hammer in the period) and half-cock the musket, before pouring the powder and ball down the barrel, and ramming the remnants of the paper cartridge home with his wooden or metal ramrod. Having firmly seated the charge, he would then cock the musket, taking the weapon from safety to fire mode. Taking aim, he would pull trigger, sending the piece of flint held in the jaws of the musket's cock spinning towards the frizzen (hammer). The stone striking the frizzen would produce a shower of sparks, and simultaneously expose the powder in the pan to those sparks. The resulting explosion would jump from the pan, through the touchhole to the main charge, and propel the ball out the end of the musket towards its intended target. The twin explosions in the pan and barrel produced a unique double sound when muskets

---

6   National Records of Scotland: 2950/4/752, Dunvegan Castle, letter from Col McLeod to the Col of the 73rd, 1787.
7   See for example, British soldiers picking off individual American officers and artillerymen at Eutaw Springs at that range, chapter nine below.

are fired individually. The noise was much less of a singular *bang* and more of a rapid *fuh-toof*. Having clothed and armed our soldier, we should now send him into campaign and battle.

The complex manoeuvres of drill were designed to teach the soldiers how to march quickly together as a body, to prepare them for particular motions on the battlefield, and above all, how to load and fire their flintlock weapons quickly using muscle memory. Soldiers would often comment that combat bore little resemblance to these complex drills, but that loading and firing their weapons repetitiously in peacetime was vital to ensure survival in actual combat. The process of actually loading and firing the weapon was referred to as the exercise *handgriffe*, whereas more complex manoeuvres and motions with large bodies of troops had more specialized terms. Officers took this work seriously, and would punish individual men, units, and more junior officers if errors were committed in the process of these manoeuvres.

Daily work depended greatly on context and environment. Usually, on campaign a large portion of the mess group would be assigned to manual labour, such as cutting wood or digging and hauling earth. One member of the mess group might be detailed for food preparation. The army frequently sought out soldiers with particularly useful trades such as cobblers, tailors, carpenters, masons (stoneworkers), and bricklayers.[8] Soldiers were often paid slightly more for work details. Unlike peacetime garrison duty, there was relatively little 'free time' in the campaign setting. Prussian soldier Ulrich Bräker described the nature of daily work, giving a blow by blow of the *Kameradschaft*'s duties:

> One man cleaned his musket, another did laundry, the third cooked, the fourth mended breeches, the fifth [repaired] shoes, the sixth cut wood... each tent had its six men and one extra, among these seven, one always had to be clean and prepared. Of the six remaining, one went on guard, one cooked, one fetched provisions, one gathered wood, one went after straw, one handled paperwork.[9]

There were, of course, exceptions to this. If the army was granted a rest day after a long march, the men would be relatively free to move about the camp, provided they did not attempt to leave the picket lines. Once again, Bräker gives us an excellent account of this type of freedom outside Pirna in 1756: 'With exception of the watch, everyone could do as he pleased, bowling, horseplay, in and around the camp.'[10] Fires were often extinguished at sunset, and a special squad of guards was detailed to be sure that noise was kept to a minimum after sundown. In the course of a siege, the daily workload of a soldier, especially in the besieged fortress, would greatly increase. Officers noted the tiring natures of sieges and all the 'various chores necessitated by

---

8   Alexander Campbell, *The Royal American Regiment: An Atlantic Microcosm* (Norman: University of Oklahoma Press, 2010), p.136.
9   Ulrich Bräker, *Lebensgeschichte und Natürliche Abentheuer eines Armen Mannes von Tockenburg* (Zurich: Hans Heinrich Füssli, 1789), p.143.
10 Bräker, *Lebensgeschichte*, p.143.

siege operations.'¹¹ In the final stages of a siege operation, it was common for the defenders to go without sleep entirely. Soldiers would work in various states of dress. During the 1759 siege of Fort Niagara, a French sortie encountered a number of British soldiers who were 'naked to the waist.'¹² A battle would interrupt the normal rhythms of army life in a number of ways.

Peacetime garrison, by contrast, was a more flexible environment. After officers called the roll and performed drill if required, soldiers were free to spend their afternoons working in the civilian sector or taking their ease. Depending on the army and year, soldiers might have lived in a fortress or barrack room or have been quartered on the civilian population. Civilians and officers complained that soldiers grew idle and fat in peacetime service:

> If the peace continues very long, I may live to see the foot [infantry] of England carried in waggons from quarter to quarter, for with their vast size and the idleness they live in, I'm sure they can't march... soon [a soldier] is incapable of wielding anything more than his musket. His hands become as delicate as a young girl and are no longer equal to gripping a spade or pick.¹³

This polemic author seems to be exaggerating. Other observers took a decidedly different view. In late-eighteenth-century Prussia, a traveller found off-duty soldiers 'without uniform of any kind, dirty, uncombed, some even without their breeches, going about just as they pleased. There are soldiers on every street corner, pursuing every means of employment imaginable.'¹⁴ Ulrich Bräker portrays a quite similar scene in the 1750s: 'hundreds of soldiers occupied themselves loaded and unloading merchant's wares, while the timber-yards were full of toiling warriors.'¹⁵ In conclusion, whether or not the army was on campaign, the life of soldiers could be quite physically laborious. Unless on a rest day, or during an afternoon lull in fortress garrison service, eighteenth-century soldiers worked industriously at the business of war or contributed a ready workforce to the economy.

Soldiers serving far from home in a colonial environment might be called to endure even more trying hardships than in Europe. A German-speaking soldier with Burgoyne's British army in North America commented:

> The banks of the lake are covered with the thickest woods, and every time a camp had to be pitched, trees had to be cut down and the place cleared. In spite of the hard work, no other provisions were furnished than salt meat and flour. As each soldier had to bake his own bread, and no ovens for

---

11   Pierre Pouchot, Michael Cardy (trans.), Brian Leigh Dunnigan (ed.), *Memoirs of the Late War in North America between France and England* (Youngstown: Old Fort Niagara Association, 2004), p.211
12   Pouchot, *Memoirs of the Late War*, p.234.
13   Percy Sumner, 'General Hawley's Chaos,' *Journal of the Society for Army Historical Research*, vol.26, no.107, (Autumn, 1948), p.93.
14   Jacques-Antoine-Hippolyte, comte de Guibert, *Journal D'un Voyage En Allemagne* (Paris: Würtz, 1803), p.166.
15   Bräker, *Lebensgeschichte*, p.121.

baking the same were there, he had to either bake it in hot ashes or on hot stones. This bread was, of course, very hard and heavy, and required good teeth. Furthermore, there was neither whisky nor tobacco, which the German soldiers were accustomed to have. I consider these last indispensable for soldiers... It is not my intention to pity the soldier. He cannot always find things as he is accustomed to having them. He must know how to endure the hardships of his profession without murmuring. However, it would be better to prepare him rather than have him come upon these hardships unexpectedly.[16]

In addition to hardships generated by tropical heat in the Caribbean, India, or Africa, or the freezing weather of Canada, soldiers often faced severe shortages of food, clothing, and equipment when deployed far from their country of origin. Disease, introduced to men who acted as the front line of settler colonialism, often had fatal consequences during the infancy of scientific medicine. Finally, when deployed abroad in hostile environments, soldiers faced the final hardship of being separated from their families and local communities, and for common soldiers, the postal systems employed by imperial states were rudimentary even in the late eighteenth century.

The world of soldiers was far less glamorous than is often depicted on the screen or canvas. Despite the fact that these men formulated narratives about the importance of combat to their lives, explicitly focusing on those narratives can distort the historical framework by making soldiers seem overly distinct from the civilian communities from which they were drawn. When these men volunteered, were conscripted, or pressed into military service, their worlds changed fundamentally and forever. Despite that, before resocialization became the norm for military training, certain eighteenth-century soldiers remained closely tied with their former identities via emotive and familial bonds, and laboured much as they did in civilian life.[17]

The Experience of Battle

For infantrymen between 1733 and 1783, the experience of battle began like any other army day. Awaking in camp or in the open lying under arms, the men would receive orders to prepare for the day. An army planning to move out in the early morning hours might give their soldiers a choice of whether or not they wanted to pitch tents or stay in the open.[18] A wet night in the open might lead to complaints

---

16 August Wilhelm du Roi, *Journal of Du Roi the Elder* (Philadelphia: University of Pennsylvania Press, 1911), pp.90–91.
17 Resocialization is the process of eroding the identities of the individual in order to prepare that individual for a new type of life. In context of military training, soldiers are psychologically retrained to place the needs of the group ahead of their individual needs. Soldiers go through this process voluntarily in the twenty-first century, the detainees of prisons and mental hospitals are compulsorily subjected to it.
18 Anon., *Offizier-Lesebuch, Historisch-Militärischen Inhalts, mit untermischten interessanten Anekdoten, von einer Gesellschafts Militärischer Freunde* (Berlin: C. Matzdorff's Buchhandlung, 1793), pp.184–185.

from the men.[19] Time permitting, commanders might allow their men to cook and eat. Hungry or not, enlisted men often commented on this in their diaries and memoirs.[20] An attacking army might begin its march at 3–4:00 a.m., while an army in a defensive position might enjoy a more leisurely battle morning, waiting for the enemy to appear. Orders to shift position, wait for the arrival of more forces, or watch the passage of men in other units were quite common.[21] The first sight of enemy troops was often recalled by men, even those who wrote their experiences many years later. For troops approaching the enemy, they might be struck by the size of the enemy army deployed in battle array or note that the enemy was taking cover behind terrain.[22] For troops standing on the defensive, the first sign of trouble would be the appearance of the heads of enemy columns of troops.[23]

At this point, the defenders would have time to observe the enemy movements, and the attacking force would begin the process of deploying from columns into line of battle. Columns of platoons would shake out into the three-four deep linear battle array. This process might take a number of hours, although certain armies could accomplish it with worrying speed. Enlisted men might recall which regiments they stood by in the line: their friends collected into brigades.[24] The attacker having deployed into linear array, the advance in line began, with officers, both on foot and on horseback, trying to maintain the alignment of formations relative to terrain and other units. This was in and of itself a challenging process. Then, the fighting began.

Regardless of their pace, the enemy seemed to come closer with frightening rapidity. Being targeted by enemy artillery was always harrowing, even if the guns themselves were more likely to be smaller than the iconic 12- or 24-pounder guns of the train. Men recalled in detail which type of anti-personnel rounds the enemy might be using. Roundshot, or cannonballs, might be relatively harmless in sheer numerical terms, but seeing a man you knew hit by a 6- or 12-pound cannonball could be intensely demoralizing. Fire from artillery tearing through the forest above an infantry position could cause showers of splinters and branches to reign down on the men. Canister or grapeshot, a collection of smaller anti-personnel balls in a container, might be used at relatively long range: at between 500 and 300 yards, the enemy would start to open fire with this sort of ammunition. The psychological pressure to return fire at this point would have been enormous, even though the enemy was still on the maximum end of small arms range.

The men received some psychological relief when, under orders or their own initiative, they began to fire at the enemy. The enemy would almost invariably

---

19 Christian F. Zander, *Fundstücke – Dokumente Und Briefe Einer Preußischen Bauernfamilie: (1747–1953)* (Hamburg: Kovacĭ, 2015), p.62.
20 William Todd, *The Journal of Corporal William Todd* (Stroud: Sutton Publishing Limited for the Army Records Society, 2001), p.168; Johann J. Dominicus, *Aus dem Siebenjährige Krieg* (Munich: C.B. Beck, 1891), pp.61, 66.
21 Anon., *Offizier-Lesebuch*, pp.185; Todd, *Journal of Corporal William Todd*, p.164.
22 Anon., *Offizier-Lesebuch*, p.185; Johann J. Dominicus, *Aus dem Siebenjährige Krieg* (Munich:C.B. Beck'sche Verlag, 1891), p.61
23 Todd, *Journal of Corporal William Todd*, p.164.
24 Todd, *Journal of Corporal William Todd*, p.165; Anon., *Offizier-Lesebuch*, p.185.

return their fire. As a result, particularly when conducted at ranges of over 100 yards, these firefights might take some time. Depending on the sort of terrain the fighting took place in, linear formations might break down, and men might seek cover and continue to fire on their own initiative. A battalion or brigade, pressed by enemy forces, might retire upon the second line of troops and be encouraged by the firmness that these fresh troops displayed. Appearances by well-known officers, generals, or even monarchs rarely rallied fleeing troops, but could greatly encourage those still willing to fight. Sometimes, during lulls in the fighting, officers exhorted their men not to fire uncontrollably, but only when the officers ordered it. Men could be equally discouraged by appearance of fresh enemy forces. Soldiers would often report using most or all of their ammunition, and they frequently carried between 30 and 60 rounds. Many soldiers reported the process of running for more ammunition collected in rear areas, taking ammunition from fellow soldiers, or even stripping enemy ammunition carts and supplies that fell into their hands.

Eventually, one side or the other would withdraw, sometimes fleeing the battlefield in haste, pursued by enemy light cavalry. In particularly intense combats, men might lay on their arms during the night, waiting to see if the enemy had withdrawn during the darkness. In exceptional cases, the fighting might continue the next morning, and sporadic fighting might continue after darkness. Infantrymen might pursue the enemy some distance, before usually being deterred by the enemy rearguard.

Men vividly recalled the cries of the dying and wounded, and details of the wounds that they, their friends, and even enemy soldiers suffered. The aftermath of battle usually involved the surviving infantrymen desperately trying to move wounded men to hospital stations in the rear areas. Soldiers particularly remembered how the cries of the wounded men would intensify upon being moved. The horror and location of the gruesome wounds produced by large-calibre firearms was recalled with detail by the men who took part. Soldiers who had made themselves scarce during the action would begin to reappear, often facing reprimands from their comrades, but rarely real punishments. Some men used the confusion of battle as an opportunity to slip away and desert from the army. The vast majority of men who remained might be ordered to complete the normal military work in the aftermath of the battle, but by this point, many were easily distracted by the idea or presence of food and set about trying to satisfy their rumbling stomachs. During the battle, men recalled drinking any water that was available, even if dirty. After the battle was done, food, and potentially alcohol, became primary objectives for the troops. Some recalled feeling depressed when the battle was done, especially when defeated. Often, their last memory of the day of battle was falling asleep to the sounds of their comrades foraging and preparing food.

The Battalion

Upon enlisting, infantrymen were assigned to a regiment, which in the eighteenth century was the common organizational unit for all types of units: infantry, cavalry, and artillery. Those organizational units were further subdivided into tactical units:

battalions for infantry, squadrons for cavalry, and batteries for artillery. This book is primarily concerned with infantry, so as a result we will focus on the battalion as a unit on the battlefield.

Battalions were the standard tactical unit across the armies of military Europe. Many armies – such as the Austrians, French, Prussian, and Russian – divided large organizational regiments into multiple tactical battalions. A Russian, French, or Austrian regiment might have two to four battalions, while a Prussian, Spanish, or Swedish regiment had two or very occasionally three battalions. As a result, an organizational unit that might contain over 1,500 men was usually tactically sorted into groups of between 500 and 700 men. The British and American armies in the eighteenth century were an outlier, as each regiment usually only possessed one battalion. There were exceptions to this but, overwhelmingly, one regiment equated to one battalion in the Anglophone world. The Hanoverian troops who fought alongside the British for much of this period followed continental rather than English practice in battalion numbers. Thus, while in the Anglophone world, a regiment and a battalion were relatively equivalent, across most of military Europe, military men spoke of battalions when discussing tactical formations of infantry on the battlefield.

Battalions were themselves subdivided into a number of small tactical units. Here, European military men used a bewildering array of terms, but divisions, grand divisions, and, above all, platoons were commonly used in most armies. Companies were, at this time, only an administrative sub-unit of a battalion. The number of platoons in a particular battalion varied between different armies. By the 1750s many armies had adopted a standard of eight platoons of their own determination or adopting Prussian practice. This process was not even. In the same army, different battalions might have a differing number of platoons based on the preference of the colonel-proprietor, *chef,* or *inhaber*. In the American War of Independence, Prussian drillmaster Friedrich Wilhelm, Freiherr de Steuben (Baron Steuben), was horrified to find that:

> With regard to military discipline, I may safely say that no such thing existed. In the first place there was no regular formation. A so-called regiment [Steuben here references the Anglophone outlier] was formed of three platoons, another of five, eight, nine, and the Canadian regiment of twenty-one. The formations of the regiments was as varied as their mode of drill, which only consisted of the manual exercise. Each colonel had a system of his own, one according to the English, the other according to the Prussian or French.[25]

Steuben's anxiety was alleviated when he managed to institute four divisions and eight platoons as standard across all American battalions.[26] In the Russian army of the middle of the century, the exact organization in a particular battalion could

---

25 Friederich Kapp, *The Life of Frederick William von Steuben* (New York: Mason Brothers, 1859), p.118.
26 Kapp, *Steuben*, p.200.

differ depending on whether both battalions of a regiment were deployed with the army. A British officer observing the Russians reported to the Duke of Cumberland:

> The Russian Infantry draw up four deep; except when they fire, and then only three deep, and the front rank only fix their bayonets. Their battalions are generally told off into sixteen platoons besides the grenadiers and their fire begins upon the right and left, and ends in the center ... When two battalions are joined they are divided into four grand-divisions, each battalion into two divisions of eight platoons each besides the grenadiers, and then the 1st and 5th, and 2nd and 6th, 3rd and 7th and 4th and 8th platoons from right and left of each grand division always fire together.[27]

In whatever number of divisions and platoons a hypothetical battalion happened to be divided, it was composed of infantrymen who had to be drilled. These infantrymen were overwhelmingly drawn from rural villages and had to be carefully instructed in order to operate as a unit on the battlefield. The preferred method across military Europe was to start with the individual soldier and progress into larger and larger units until the entire battalion could function together.

Taking Prussia as an example, a new recruit would be assigned into the charge of a non-commissioned officer, who would proceed to train him to walk with the military step, follow the manual of arms, and load and fire his musket. The comte de Guibert described the subsequent steps:

> When he is fully instructed in these matters he is added to the platoon of recruits. He is examined daily, by a company officer who ensures that he has been properly trained up to this point. The platoon of recruits is drilled daily, a captain inspects it every day, and decides the men who have learned the most to incorporate them into the [regular] platoons... then they do drills and manoeuvres with the battalion.[28]

When battalions were poorly trained, this process could be reversed. This comes to us clearly from the example of the Prinz Friedrich Regiment of Braunschweig-Wolfenbüttel, during its deployment to Canada during the American War of Independence. The overall commander of the Braunschweiger force, Adolph Friedrich von Riedesel, provided a clear outline of how training a deficient unit was to proceed.

In a scathing critique, Riedesel instructed *Obristleutnant* Christian Julius Prätorius, the commander of the Prinz Friedrich Regiment, on how to retrain his unit. He believed that Prätorius had drawn his regiment together too early; the individual companies lacked sufficient training. Riedesel informed Prätorius, 'True, the companies have put in the usual amount of drills, but the faults have not been noted,

---

27 Royal Archives (RA): Cumberland Papers Main Series 43, f.5b, 'Dress of the Russian Foot, 1749'.
28 Jacques-Antoine-Hippolyte, comte de Guibert, *Observations sur la Constitution Militaire* (Berlin: publisher unknown, 1778), p.111.

and every exercise has not been corrected until the men did it as it should be done. You have drawn the battalion together too early, as this ought not to have been done until all the companies were equally efficient.'[29] So, in a situation rather the opposite of the American Army during the same period, Riedesel argued that training should be carried out on a smaller scale before occurring on the regimental level.

Riedesel sadly noted, 'You will have to commence again from the beginning, first in files and then in companies, to march, do manual exercises, load, and repeat this until all the companies are equally well drilled.' He exhorted Prätorius to 'make the maladroit and the ignorant step forward every time,' for punishment. The letter made clear that Riedesel no longer trusted Prätorius. He appointed another *Obristleutnant*, Friedrich Baum of the Braunschweig Dragoon Regiment, to oversee the drilling of Regiment Prinz Friedrich. 'Baum ... has received orders from me to see them drill frequently, and to tell you when he finds the companies so far advanced that you can draw the battalion together, and then you will unite all the companies.'[30]

The Prusso-American drillmaster Steuben also noted that the preferred method of training troops was to start small and work their way larger, as he noted in a letter to Freiherr de Gaudi in 1785:' Judge then, whether I could amuse myself with the management of arms and parades. Contrary to my principles, I was forced to begin my task at the wrong end, and after executing great maneuvers with six or eight thousand men together, I have sent my generals and colonels to learn the manual exercise.'[31]

Formations and Tactics: Columns

Most eighteenth-century soldiers spent the vast majority of their time in two general formations: columns of march and lines of battle. Columns were used to facilitate moving large bodies of troops to and from the battlefield; lines were the most commonly employed tactical formation in combat. Columns were deep formations, between four and 25 men wide and between 25 and 100 deep.[32] They were designed to snake after the leading element, making them ideal for quick manoeuvres when outside of battle. When on route marches over long distances, most armies allowed their men to walk at their own pace, without worrying about the keeping in step or cadence. Columns were occasionally used as a battle formation, usually in an ad-hoc way.[33]

---

29 'Correspondence of General Riedesel,' in *Hessian Documents of the American Revolution* (Boston: G.K. Hall, 1989), p.HZ-1 932.
30 'Correspondence of General Riedesel,' in *Hessian Documents of the American Revolution* (Boston: G.K. Hall, 1989), p.HZ-1 932.
31 Kapp, *Steuben*, p.700.
32 British Library (BL), King's Manuscripts 241, f.12. 'Remarks on the Prussian Troops and their Movements'.
33 See Chapter 8, below.

On the march, commanders preferred to keep their columns 'short and fat.'[34] In practice this meant that the width of the marching column was not dictated by the width of the road or path it was travelling on, but rather the tactical subdivisions of the unit. Thus, the Austrians preferred to march in divisions (an eighth of the battalion) and the Prussians by platoons (again, an eighth of the battalion).[35] With the American Continentals, the frontage of the columns varied. Early in the war, they marched in relatively narrow columns of a few files, but after the appearance of Freiherr de Steuben as inspector general of the army in 1778, the main body of Continental troops appear to have primarily marched with a platoon frontage, following Prussian practice.

As early as 10 April 1778, George Washington noted in his general orders: 'as marching men by files has an unmilitary appearance and a tendency to make them march in an unsoldierlike manner, all parties commanded by commissioned officers are to be marched by divisions.'[36] This command to march by divisions was superseded by an order to march by platoons on 1 June, and although the Continental Army continued to march by files when particular circumstances required, even when advancing through Indian country – where the main body of the army seems to have marched in columns of platoons – at least 10 files wide, and usually two deep.[37] These platoons would have been stacked up eight deep per battalion in the army. Armies would march with or without intervals between the platoons, in open or closed columns. Having reached the battlefield, officers would begin the time-consuming process of deploying or displaying the line.

Deploying and Displaying

There were essentially two methods of transforming from a march column to a line of battle: the parallel march or the *deployiren*. The parallel march was by far the most common method of deployment utilized by the armies of the period. In order to use this system, two ingredients were required. First, the deploying army had to march with its columns in open platoons (with intervals between the sections). This, vitally, would allow the space necessary to form the line of battle when the troops were ordered to quarter-wheel into line. The second was that the army needed to march parallel to the enemy in order to form the line of battle facing their opponent. The troops could approach the enemy for a time head-on but needed to swing into a parallel march in order to form the line of battle facing the enemy. Positioning the

---

34   Duffy, *Instrument of War*, p.389.
35   Duffy, *Instrument of War*, p.389; Duffy, *The Army of Frederick the Great* (2020), p.138.
36   Library of Congress (LOC): Washington Papers, General Orders, 10 April, 1778.
37   LOC: Washington Papers, General Orders, 1 June, 1778; Frederick Cook, *Journals of the Military Expedition of Major General John Sullivan* (Auburn: Knapp, 1887), pp.26, 68, 99, 146, 150; Anon., *Proceedings of a General Court Martial… for the Trial of Major-General Lee* (New York: Unknown, 1864), pp.36, 50, 146, 147. According to Steuben's *Regulations for the order and discipline of the Troops of the United States,* a platoon had to consist of at least 10 files.

entire army was the most time-consuming part of this process; the actual quarter wheel into line of battle could be achieved with frightening rapidity.

The second form of deployment was the *deployiren,* first developed in Prussia and copied by most major states. In the 1740s, the Prussian Army under Frederick the Great developed a new sort of deployment method. Rather than approaching the battlefield parallel to the enemy front, battalions would march directly at the enemy, and then the lead platoon would slow its march, while the following platoons would race out from behind the lead platoon to the right or left sides of the formation, quickly forming a line of battle. This was a complex and innovative manoeuvre, which the Prussians practiced on the battlefield a couple of times in the War of Austrian Succession and the Seven Years War, and Frederick of Prussia termed it the *deployiren.*[38] By the 1770s, although not officially adopted by any army, the manoeuvre was relatively common knowledge in Europe, and before Steuben's Regulations, there is evidence to suggest that Continental units employed it at Germantown.[39]

For a time, this Prussian development had the potential to overawe its opponents, as Pastor Täge recorded at Zorndorf:

> Never will I forget this approach. Majestically beautiful, and in still, quiet order, the Prussians advanced. Suddenly – and here I wish that the reader could truly feel this terrifyingly beautiful view – suddenly the Prussians deployed, and henceforth presented a long line in oblique order. The Russians themselves were astounded by this unwonted sight, which as is generally maintained, was a great triumph of the great king's tactic of that time.[40]

By the 1770s, most major armies were at least theoretically capable of deploying out of columns using the *deployiren.* However, armies continued to form the line by the parallel march. American Colonel Israel Shreve described this action in his journal entry for the Battle of Short Hills on 26 June 1777: 'we marched in a column [and] wheeled to the right to form line of battle.'[41] Christopher Duffy asserted that the Prussians used the *deployiren* sparingly in actual combat: Soor in 1745; Lobositz, Reichenberg, and Gross-Jägersdorf in 1757; and Torgau in 1760.[42] That the tactic was successfully executed by subordinate commanders (Bevern at Reichenberg and Lehwaldt at Gross-Jägersdorf) indicates that there was nothing distinctly personal to Frederick in the usage of this tactic and its transmission to other states.

---

38 For a discussion of Prussian deployment systems, see: Christopher Duffy, *The Army of Frederick the Great,* 2nd Edition (Chicago: Emperor's Press, 1994), pp.122–126.
39 Enoch Anderson, *Personal recollections of Captain Enoch Anderson* (Wilmington: Historical Society of Delaware, 1896) 45. Captain John Markland indicates that his men were formed in a 'close column' which was utilized during this manoeuvre. John Markland, 'Revolutionary Services of Captain John Markland,' *The Pennsylvania Magazine of History and Biography,* vol.9, no.1 (1885), p.107.
40 Adam L. Storring, 'Pastor Täge's Account' in Burns (ed.), *The Changing Face of Old Regime Warfare,* p.218.
41 Louisiana Technical University, Prescott Memorial Library: Buxton Collection, Israel Shreve Papers, Colonel Israel Shreve journal entry for 26 June 1777.
42 Duffy, *The Army of Frederick the Great* (2020), p.143.

Formations: Line of Battle

In viewing the modern representations of linear warfare in reenactments, wargames, and film, it is easy to believe that these formations were rectangular blocks of men, certainly wider than they were deep – but still blocks. Instead, we must imagine this as the age of *linear* warfare, where armies fought in lines of battle exponentially longer than they were deep. This can be observed of battalions as depicted in period artwork, such as David Morier's *Review of the Norfolk Militia*. The three-deep formation he depicts is over 100 files long. The officers faced immense challenges in keeping such a long formation in order, or even in contact with itself.

In the minds of the public, the robotic eighteenth-century soldier marches in permanently arranged files of two or three deep, but in reality, they altered their depth as required by need:

> Apart from the customary form of formation in three ranks, (for the old practice of putting them on four had been almost continuously abolished before the First Silesian war), the Prussian infantry and cavalry had also been trained in two ranks; this manoeuvre took place especially in the Seven Years' War years, when the Prussians, in order to hide their weakness, often marched on only two ranks; indeed, in the camp near Strehla in 1760, General Hülsen was forced to place 5 battalions from his left wing... only one rank deep, in order to cover the whole front of this position...[43]

As was so often the case in eighteenth-century warfare, even Prussian officers were flexible enough to deviate from the drill manual in order to match their formations to the needs of the moment. With the realization that officers could change the depth of their formations as needed, what was standard practice? At the start of the period, it was common to form for battle four men deep. The Prussians made the transition to a three-deep line between November of 1740 and June of 1742.[44] In 1749, an observer noted, 'The Russian infantry draw up four deep; except when they fire, and then only three deep.'[45] This likely indicates a period of transition was underway in the Russian Army in the era between the War of Austrian Succession and Seven Years War. The French Army reduced to three ranks in the same period.[46] The Austrians made a similar transition early in the Seven Years War.[47] The British fought in a three-deep formation throughout the Seven Years War but transitioned to an open-order, two-rank line in 1775 during the American War of Independence.[48] The Continental Army, as well as

---

43 Franz Ludwig von Haller, *Militärischer Charakter und merkwürdige Kriegsthaten Friedrich des Einzigen*, (Berlin: Demigke, 1796), p.12.
44 Duffy, *The Army of Frederick the Great* (2020), p.134.
45 RA, Cumberland Papers Main Series 43, f.5b. 'Dress of the Russian Foot, 1749'.
46 Christopher Duffy, *Military Experience in the Age of Reason*, p.110.
47 Duffy, *Instrument of War*, p.441.
48 Matthew H. Spring, *With Zeal and With Bayonets Only: The British Army on Campaign in North America*, (Norman: University of Oklahoma Press, 2008), pp.139–142.

the French in North America during the American War, also adopted this two-rank formation.[49] Hessian forces were very hesitant to join the British in an open-order line, but did form two deep, altering from the standard three ranks they had used during the Seven Years War.[50]

Thus, while a two-rank formation was more unusual in Europe, it was the standard in North America. In drill books and regulations, the three-rank formation was the hypothetical standard, but if the Americans, British, French, and Hessians formed in two ranks during the American War of Independence, and the Prussians did so for much of the Seven Years War, there was certainly plenty of opportunity to find infantry arrayed in two ranks, even if a three-rank formation was statistically more normal throughout the period as a whole.

With the battalions formed in their ranks, deployed in line facing the enemy, the time had come to finally engage. There was a bewildering array of tactical options available to the eighteenth-century battalion commander. Essentially, these fell into two categories: to intimidate the opposing infantry into giving ground, or to cause the enemy to retreat as a result of casualties delivered by firepower.

Tactics: Cadenced Marching

The re-emergence of cadence marching, whatever the exact lineage of the practice, was a development with deep tactical implications. This cadenced marching may be the dominant image of eighteenth-century warfare that resonates with the public today: Michael Leggiere's description of eighteenth-century infantry grounds itself in the importance of cadenced marching, or at least seems to in describing it as a 'rigid, close-order, closely supervised, cohesive, slow, methodical, unitary organism.'[51] However, as Christopher Duffy has argued since the 1980s, cadenced marching primarily impacted battles *before* the shooting started. In 1987, he concluded, 'It is, however, much more difficult to establish if cadenced step survived very long on the conditions of the battlefield... among infantry, the step was very easily lost through the effects of casualties, or obstructions like stones, holes, or a ploughed field.'[52] Discussing the complex steps and manoeuvres of the drill-square, the Prussian officer Lossow comments: 'This was hardly used at all in war, as far as I know... on the battlefield... one had to break up the orderly lines, because one was

---

49　Marquis de Lafayette and Stanley J. Idzerda, *Lafayette in the Age of the American Revolution: Selection Letters and Papers* (Ithaca: Cornell University Press, 1977), vol.1, p.91; Henry Clinton and William B. Willcox, *The American Rebellion: Sir Henry Clinton's Narrative of His Campaigns, 1775–1782* (New Haven: Yale University Press, 1954), p.95; Ludwig von Closen and Evelyn M. Acomb, *The Revolutionary War Journal of Baron Ludwig von Closen, 1780-1783* (Chapel Hill: University of North Carolina Press, 1958), p.92.
50　Rodney Atwood, *The Hessians: Mercenaries from Hessen-Kassel in the American Revolution* (Cambridge: Cambridge University Press, 1980), pp.61, 66–67, 82, 242.
51　Michael V. Leggiere, 'Napoleon and the Strategy of the Single Point,' in Hal Brands, *The New Makers of Modern Strategy* (Princeton: Princeton University Press, 2023), pp.320–321.
52　Duffy, *Military Experience*, pp.112, 203.

forced to march through terrain that would have rarely facilitated this over long distances.'⁵³

Georg Heinrich von Berenhorst, an aide to Frederick of Prussia, was blunter about the cadenced step in the actual advance to contact. He noted:

> The advance against the enemy is the most vital tactical task… the step is the basically building block of all our fancy tactical evolutions… with lowered sarissa, on open, leveled ground, Greek phalangites advanced step by step. The Spartans alone accompanied their marches with flutes and music. If you so desired, you could do the same with our cadenced marching today. But do not forget: you are not Spartans, and our world reeks of gunpowder…When General Saldern is able to speak with Gustavus Adolphus and Montecuccoli in the afterlife, he can ask them: was the entire earth a flat plain in your day? … the battalion of two hundred or two hundred and fifty files makes a fine impression as it advances on a wide front towards the spectators… But these splendid evolutions are only the happy vision of the exercise field, and they do not always work even there. A ploughed field or tangled meadow is a powerful hinderance: harmony turns to dissonance… the front falters.⁵⁴

None of this means, however, that the cadenced step failed to decisively impact the eighteenth-century art of war. Instead of being a prominent feature on the battlefield itself, cadenced marching allowed officers to retain vital control and synchronization in complex pre-battle manoeuvres before and during the formation of the line. Marching in step facilitated synchronization when an entire brigade, wing, or even army needed to quickly deploy using either the processional or *deployiren* methods. The Prussian skill in cadenced marching helps explain, to some extent, how overawed non-Prussian armies were by the pre-battle movements of the Prussian Army: it is what Täge was referencing above with his comment regarding the 'great triumph of the great king's tactic of that time.'⁵⁵

Tactics: Intimidation

The tactics of intimidation relied on two principal threats: first, the threat of unfired muskets being brought closer and closer to their target; and second, the threat of dying a gruesome and messy death to bladed weapons in hand-to-hand combat. Both of these factors played a powerful role in terrifying enemy soldiers. Officers believed that if they could bring one or both of these factors to bear on the mind of their adversaries in a direct enough way, the enemy forces would flee from action.

---

53   Ludwig von Lossow, *Denkwürdigkeiten Zur Charakteristik Der Preussischen Armee Unter Dem Grossen König Friedrich Dem Zweiten* (Glogau: Carl Heymann, 1826), p.242.
54   Georg Heinrich von Berenhorst, *Betrachtungen über die Kriegskunst* (Leipzig: Fleischer, 1798), vol.2, pp.423–424.
55   Storring, 'Pastor Täge's Account', p.218.

Advancing to under 50 yards of an enemy's position with unfired muskets was a difficult prospect for the men marching forward into a hail of enemy fire, but almost as harrowing for the men firing at them, who knew that their deadly first volley would be delivered at closer and closer range as they advanced. Famously, the Swedish Caroliner of the Great Northern War had played on both of these fears, charging into close range with loaded weapons, firing, and then continuing to advance with sword, bayonet, and pike in hand.

A favoured tactic of many eighteenth-century commanders was to advance without firing, bringing the battalion ever closer to the enemy position.[56] With the technological weapon systems available at the time, firepower was exponentially more accurate under 50 yards. In a treatise that drew experience from the Seven Years War, Major General Richard Lambert, the Earl of Cavan, explained the advantages of reserving fire, even if the enemy was firing on your battalion:

> In advancing to this or any other attack, the men are to march briskly on, and for these essential purposes. First, that it serves to animate them the more; and next, that it proves the only means to strike dismay and confusion into the enemy, and to induce him in consequence to throw away his fire, either precipitately or at too great a distance.[57]

The very act of advancing on the enemy with loaded muskets would, in Cavan's view, encourage your men, intimidate the enemy, and cause them to fire too soon. By firing too soon, their position would be even further eroded in the face of your bayonets. In his view, the infantry force which held their fire the longest would always emerge victorious. He explained this at length:

> As the enemy will be ignorant of your design to reserve your fire, and to depend solely on your bayonet, he will most probably give his fire on your advancing, which as surely as he does, his is inevitably lost; there being, I may venture to affirm, not an instance to be produced, where soldiers have imprudently thrown away their fire, and not finding the effect which they hoped to receive from it, did not, on being instantly after close pushed, think themselves incapable of all further resistance, and in consequence turn round on those in rear to seek flight that way and by the panic they communicate, throw all into confusion and disorder.[58]

As Cavan quite honestly noted in the next sentence, and what we have already seen at from Johann Macquire's report on the Battle of Guastalla, there was only one potential flaw with this plan: 'The only danger attending an attack of this sort is, the risk you run of your men giving their fire first, and then the tables may be easily

---

56 For a full discussion of this tactical philosophy, see chapter four.
57 Richard Lambert, Earl of Cavan, *New System of Military Discipline Founded upon Principle by a General Officer* (Philadelphia: Aitken, 1776), p.188.
58 Cavan, *New System of Military Discipline*, p.151

turned upon yourself.'⁵⁹ Cavan then encouraged officers to not permit their front ranks to load. By reserving fire, Cavan believed, one battalion would overcome its opponent which fired first.

Far from being a merely hypothetical exercise, this sort of thinking did indeed impact the battlefield. At the Battle of Germantown in 1777 and Eutaw Springs in 1781, American troops advanced on British positions with their weapons shouldered, or trailed, without firing. The advance of troops with shouldered muskets was enough to cause the enemy to retreat. American Major General John Sullivan reported, 'we drove their left wing near three miles, a great part of the time [with] shouldered arms and charged bayonets.'⁶⁰ Clarifying, Sullivan expanded: 'we were soon after met by the left wing of the British Army when a severe conflict ensued but our men being ordered to march up with shouldered arms they obeyed without hesitation and the enemy retired.'⁶¹ The threat of intimidation from troops with shouldered muskets was real, even for veteran troops like the British regulars at Germantown.

The second sort of intimidation was offered by the *armes blanches*, or cold steel itself: the understandable desire on the part of troops to avoid close combat.⁶² The terror and speed on an enemy attack close combat weapons was communicated by American pensioner Solomon Freer, who served in a regiment of light horse: 'the enemy came upon us suddenly with drawn swords and bayonets, and charged before a line could be formed.'⁶³ The veteran American officer John Eager Howard gave an indication of this sort of tactic used by the British on his men at Cowpens: 'the moment the British formed their line they shouted and made a great noise to intimidate, and rushed with bayonets upon the militia, who had not time, especially the riflemen, to fire a second time.'⁶⁴ The most frequent type of bayonet attack was one where the enemy fired – perhaps at close range, perhaps longer – and then followed that fire up with pressing into close range with bayonets. As chapter eight discusses, defending troops would usually flee rather than fight in close combat.

Tactics: Firepower

Eighteenth-century officers also had a number of options for delivering firepower to the enemy. In the Austrian army of the 1750s, officers could choose to fire as a whole battalion, by divisions, subdivisions, platoons; or they could engage in *heckenfeuer* (or *höckenfeuer*), *Retrenchment-Feuer, Doppeltes-Weeg Feuer, Gassen-Feuer,*

---

59 Cavan, *New System of Military Discipline*, p.151
60 John Sullivan and Otis G. Hammond, *Letters and Papers of Major-General John Sullivan* (Concord: New Hampshire Historical Society, 1930), vol.1, p.576.
61 Sullivan and Hammond, *Letters and Papers of Major-General John Sullivan*, vol.1, p.545.
62 This idea is explored in more detail in chapter seven.
63 National Archives of the United States (NARA): M804 Solomon Freer, Pension application of Solomon Freer W8826
64 Historical Society of Maryland (HSM): Bayard Papers, MS 109, Box 1, Item 18, John Eager Howard to John Marshall.

or *Brücken-Feuer*.[65] These methods were hardly unique to the Austrians but were debated in most major militaries. In 1755, British officer James Wolfe recommended live-fire target-practice: 'singly … by file, 1, 2, 3, or more, then by ranks, and lastly by platoons.'[66] In the 1770s, the comte de Guibert thought it was vital for French infantry to practice firing as a battalion, half-battalions, divisions, half-divisions, platoons, and half-platoons, in addition to firing by rank and firing independently.[67] In 1749, the Russians practiced firing by platoons, where 'their fire begins upon the right and left and ends in the centre', and by ranks, 'beginning with the rear and ending with the front rank.'[68] The Prussians practiced firing by platoons, two-platoon groupings (a quarter of the battalion), and whole battalion volleys, in addition to *heckenfeuer*.[69] Washington believed it was important for the Continentals to 'fire by platoons, division, and battalion standing.'[70] Thus, likely by the 1750s, and certainly by the 1760s, most major powers in military Europe practiced with similar tactical methods of delivering fire.

With the methods of delivering fire outlined above, we are left with a tantalizing question: what on earth did they all mean? In exploring the firings, we will begin with general or larger groupings of fire and move to the smaller and more specific. Thus, the natural starting place is the battalion volley.

The whole battalion volley provided officers with a number of advantages. First, it fired the maximum number of muskets available at a target, leading to saturation of fire. Second, it was relatively easy to control; officers did not have to worry about an intricate sequence of firings that could easily fall into disarray. Third, the entire battalion firing at once caused a terrifying roar which could dishearten opponents or even your own troops – George Washington's officers noted the importance of new recruits being drilled to firing as a battalion because they were unaccustomed to the noise.[71] Thus, a battalion volley provided three advantages of volume, simplicity, and intimidation.

Despite these advantages, the battalion volley possessed one major drawback: it left all muskets in the battalion unloaded for a period of perhaps 15 to 30 seconds. Although this might seem like a relatively short amount of time, in the midst of a fight between two battalions, it could be an eternity. Enemy troops moving at the quickstep (perhaps 160 paces per minute) or a run would be able to advance on your position between 50 and 100 yards. If the battalion were advancing at a jog or run and they fired on a charging enemy at 100 yards, as the British did the American War of Independence, they would be upon the firing battalion before it had a chance to fire again. This assumes that the enemy was on foot, but against cavalry squadrons charging at the trot, the calculus was even more worrying. Thus, commanders often preferred, when possible, to have smaller sections of the battalion fire, always leaving a reserve of loaded muskets to fire at the enemy if they approached into very

---

65   Duffy, *Instrument of War*, p.445.
66   Beckles Wilson, *The Life and Letters of James Wolfe* (London: Heinemann, 1909), p.255.
67   Guibert and Abel (trans.), *Guibert's General Essay on Tactics*, pp.77–78.
68   RA, Cumberland Papers Main Series 43, f.5b. 'Dress of the Russian Foot, 1749'.
69   Duffy, *The Army of Frederick the Great* (2020), pp.144–145.
70   LOC: Washington Papers, General Orders, 8 October 1779.
71   LOC: Washington Papers, Brigadier General Alexander McDougall to George Washington, 21 May 1777.

close range. This led many officers to believe that firing by platoons, divisions, ranks, or other sections of the line was more effective than firing by battalions.

Firing by sections or subsections of the battalion seemed to offer powerful advantages. First, if firing by platoons, officers believed that by the time all platoons had completed firing, the first platoon in the sequence would be reloaded. This meant a constant stream of outgoing fire towards the enemy. Second, if this was the case, that also meant that a reserve of loaded muskets, about to be fired, would be available at all times in an emergency. A well-executed fire by platoons, divisions, or subdivisions provided the prospect of a seamless rolling firing back and forth across the front of the battalion, while still providing the commander of the battalion with a reserve of muskets to use should the enemy begin a rapid advance. Thus, commanders often associated platoon fire with long-distance battles. British light cavalry officer Banastre Tarleton believed that the best way to deter an enemy attack was not to reserve fire as described above, but rather 'a successive fire of platoons or divisions, commenced at the distance of three or four hundred paces.'[72] Tarleton's American opponents seemed to agree. William Stephens Smith, a battalion commander at the Battle of Connecticut Farms in 1780 noted: 'The enemy then made their appearance upon my right in front and I order'd the two platoons upon my right to begin the fray after two or three discharges I thought the enemy appeared rather cautious in advanceing and I order'd the platoons to desist fireing and wait untill they should approach a little nearer.'[73]

For Smith, platoon firing was a way to deter enemy attack at long range, and in this particular instance, Smith believed that the result was successful enough that he wanted to allow the enemy to approach closer before he continued firing, which he then did, firing 'eight or ten rounds per man.'[74]

Firing by ranks, divisions, or other groupings of the battalion produced the same advantages as platoon firing with one notable drawback: commanders were forced to choose whether they would rather fire all of their subdivisions in succession and wait a short period to reload with no reserve of muskets, or fire slowly, maintaining a reserve of loaded muskets at the expense of a constant stream of fire. The main advantage of firing in this method was with a larger grouping of muskets, firing did more collective morale and physical damage to the enemy in each individual discharge.

The major disadvantage of firing by various sections was that this tactic, while so orderly on the peacetime drill square, tended to be very difficult to control and maintain in combat. The ability to maintain platoon firing in combat was highly sought after yet rarely achieved.[75] One of the special hallmarks of the Prussian system was supposedly the ability to maintain this platoon firing in actual combat, but the drillmaster Steuben commented: 'I have often remarked that the Prussians, after the first charge in action, no longer practice platoon firing, do not load so often in a minute and fire quite as badly as the Russians, Austrians, or French.'[76]

---

72 Banastre Tarleton, *A History of the Campaigns of 1780 and 1781* (Dublin: Exshaw and Whitestone, 1787), p.33.
73 LOC: Washington Papers, William Stephens Smith to George Washington, 10 November 1780,
74 LOC: Washington Papers, William Stephens Smith to George Washington, 10 November 1780.
75 See chapter six.
76 Kapp, *The Life of Frederick William von Steuben*, p.118.

The other, more arcane firings were usually designed for use in a specific case, rather than a general battalion-on-battalion firefight. *Heckenfeuer* was a type of skirmishing native to German Central Europe, where a small collection of two or three files would leave the main body of the battalion, fire, and then return to the battalion to reload. This provided the option to skirmish at longer range without wasting the entire fire of the battalion or robbing most of the troops of their vital opening volley. *Gassen-Feuer,* or street firing, was more common across military Europe and provided officers with options for engaging the enemy in an enclosed urban or village environment, and essentially was a return to the counter-march common to the 1500s and 1600s. After firing, a section of men would peel off, return to the rear of the formation, and be replaced by men who had loaded muskets.[77] As officers planned to engage the enemy, they did so using a tactical toolkit that largely consisted of the methods described above.

Conclusion

This chapter has explored the nature of life in the armies of military Europe, the experience of battle for common soldiers, the nature of the old regime infantry battalion, and the tactical methods that the battalion employed in combat. Although a few methods stand out as particularly favoured or developed by the Prussians, across military Europe battalions shared tactical abilities to a remarkable degree. Preferences might exist across the military of one particular state, or even a particular battalion commander, but most infantry troops practiced similar tactics across the continent and shared a mutually intelligible praxis of battle.

In his 1987 *The Military Experience of the Age of Reason*, Christopher Duffy noted the manner in which European contemporaries characterized the 'national' attributes of the 'cold' or 'hot' nations.[78] 'Cold' nations such as the Germans or Russians might be more prone to engage in battle from a distance, while 'hot' nations such as the French or Spanish might prefer to charge home with the bayonet. Despite the prevalence of this quasi-racialized thinking among contemporary officers, most European states took care to drill their soldiers using a variety of tactical practices. These officers, and even their soldiers, had the ability to select which tactic they believed to be most appropriate to a particular situation, rather than adopt a predetermined tactical selection based upon their 'national character.'

Instead of a national character, soldiers often engaged in battlefield negotiation with their officers, disobeying their orders where possible, so that they might have a better chance of surviving the battle. In order to fully understand the men who defied their officers to ran on the battlefield, fired without orders at longer ranges than their officers preferred, and avoided melee, we must understand the social world from which these soldiers were drawn. That world, the world of the eighteenth-century village, allows us to better understand these soldiers as men.

---

77 Duffy, *Instrument of War*, p.445; Duffy, *The Army of Frederick the Great* (2020), pp.145.
78 Duffy, *Military Experience*, pp.18–19.

3

Eighteenth-Century Soldiers and Their Village World

> *'He can't allow himself to be cursed as a dog's ass nor let it pass. Neither his army officer nor Herr Major von Kleist have commanded any such thing. He has enough understanding of his own that he can't allow himself to be cursed.'*[1]
>
> Deposition of Prussian Musketeer Peter Glissman of the village of Blüthen, 1770

In March of 1777, the blue-coated troops of Ansbach-Bayreuth found themselves pinned down by accurate rifle fire. When the troops had been selected to go to America, this was the last outcome they had expected. But then, they were not yet in America. Just outside the town of Ochsenfurt near Würzburg, these troops were under fire from their own rifle-armed Jäger contingent. Enlisted man Johann Conrad Döhla recalled the scene:

> Therefore, the Jäger Corps was ordered to take posts on the heights and to fire shots into the air to intimidate the fleeing rebels. But our people fired at the Jägers. In this exchange some of our troops were wounded in the legs... the exchange of shots lasted for about two hours, and because the Jägers wounded some of us, it was the cause of a great antipathy between us and them[.][2]

What had created this deadly situation? Crowded conditions on a river troop transport. Döhla seems to think the mutiny was much ado about nothing, but Stephen Popp, another enlisted man at the scene, recalled in his diary the reasoning behind the *aufruh*: 'Whoever are real soldiers get off the ships to the shore, for why should we allow ourselves to be cooped up like rogues?'[3] The situation was not fully resolved until the Markgraf Alexander, the leader of their small state, rode to the scene of

---

1 Quoted in William H. Hagen, *Ordinary Prussians: Brandenburg Junkers and Villagers, 1500-1840* (Cambridge: University of Cambridge Press, 2002), p.471.
2 Johann Conrad Döhla and Bruce Burgoyne, *A Hessian Diary of the American Revolution* (Norman: University of Oklahoma Press, 1990), pp.5–6.
3 Stephan Popp and Reinhart J. Pope, *A Hessian Soldier in the American Revolution: The Diary of Stephan Popp*, (Whitefish: Literary Licensing, LCC, 2011), p.1.

the mutiny and negotiated with the mutineers. The Markgraf arranged for two more ships to be brought up, and Alexander even accompanied them (now in four transports instead of a crowded two) for a time. This incident shows the vigour that soldiers could bring to bear when they felt unfairly treated. Mutinies were a common method of negotiating with authority figures by soldiers across all the armies of military Europe. Considering the village world that most soldiers hailed from, this is not surprising.

The eighteenth-century village world that most European soldiers hailed from was a contentious and litigious place where peasants frequently demanded their rights in court, petitioned their cases to noble and royal authority, and demonstrated a distinct lack of respect, even for those who were in authority over them. This was a common feature of life across many of the communities in military Europe. Soldiers would advocate for themselves and those close to them. Between 1733 and 1783, soldiers came from small, close-knit communities across military Europe. In almost all cases, these communities were structured hierarchically. The peasant or farming families that the majority of soldiers were drawn from had to provide taxation, service, and labour for religious and noble elites who often controlled the land they worked. However, these same elites also had specific obligations and duties to the peasantry, and the agricultural class cautiously but forcefully demanded that landed classes live up to their end of the bargain.

Eighteenth Century Soldiers' Worlds

Matthew Spring, in his groundbreaking 2008 work on the tactics of the British Army, wrote that 'the ultimate purpose of all armies is to fight, and… therefore the most fundamental task facing the military historian is arguably to study combat.'[4] Spring is correct. This specific chapter, however, focuses on examining the nature of communal links between soldiers and their communities, and takes a broader view of soldiers and their living worlds. In a book about combat, understanding the lives of the men engaged in combat is still necessary. Combat is indeed addressed, but it is not the only activity in which soldiers engaged, even if soldiers formulate combat narratives regarding its special importance in their experience.[5] As the previous chapter noted, during the eighteenth century, soldiers spent far more time drilling, marching, cooking, and labouring than they spent in battle. These aspects of army life were perhaps not as dangerous or exciting as combat, but they were difficult and arduous enough for the soldiers performing them. A majority of eighteenth-century soldiers in Britain and Prussia came from the social strata of unskilled day labourers.[6] In Britain specifically, it is possible that unemployed textile works made up the next largest

---

4 Spring, *With Zeal and With Bayonets Only*, p.xi.
5 Yuval Noah Harrari, *The Ultimate Experience: Battlefield Revelations and the Making of Modern War Culture, 1450-2000* (Houndmills: Palgrave-Macmilllan, 2008), pp.1–19.
6 Duffy, *The Army of Frederick the Great*, 3rd Edition, p.75; Glenn Steppler, *The Common Soldier in the Era of George III* (unpublished doctoral dissertation, University of Oxford, 1984), p.227.

category, of approximately 30 percent.[7] In both Britain and Prussia, men were enlisted in their early twenties, and the average soldier was approximately 30 years old.[8]

Most soldiers hailed from rural communities: small villages or hamlets. In Prussia, an average of seven villages in the Brandenburg district Stavenow at the turn of the eighteenth century gives a mean of 24 households, with a median and mode both of 27 households.[9] Likewise, in Britain, around the turn of the eighteenth century, perhaps 80 percent of the population lived in rural villages, and although it is almost impossible to estimate the average village size across Britain in the eighteenth century, most villages were in the order of 100 to 400 individuals, with the number of households per settlement apparently slightly larger in Britain than in Prussia.[10] In both Britain and Prussia, fullholder, or yeoman, households outnumbered smallholders and day-labourers in almost every village. The realities of eighteenth-century village life, as revealed by recent scholarship – such as the works of William Hagen and Peter Laslett – often challenge longstanding stereotypes regarding the nature of poverty and hardship in these rural worlds.

Households had changed greatly from the one-room medieval house. Indeed, even the poor peasants of the *Büdner* class could expect a dwelling with a separate bedroom, living room, and kitchen, with separate enclosed rooms for livestock and fodder.[11] This same process was at work in Britain, where landowners built larger houses of multiple-purpose rooms, using brick and stone, with boarded floors replacing the earth floors of the seventeenth century.[12] These household communities, '*das ganze Haus,*' looked slightly different in Britain and Prussia. In Prussia, larger multi-generational households seem to have endured to a greater extent than in Britain, where the average household size was approximately four.[13]

Between 1733 and 1783 the majority of the eighteenth-century villagers in both Britain and Prussia lived lives that were marked by greater material wealth than previous generations, and possessed a limited surplus.[14] Despite this, life expectancy was lower than modern norms. Daily work was hard, and in both Britain and Prussia the state intervened to curb the worst of poverty's effects during the century.[15] Though they were poor, peasants often possessed multiple sets of clothes,

---

7 Sylvia Frey, *The British Soldier in North America: A Social History of Military Life in the Revolutionary Period* (Austin: University of Texas Press, 1981), p.12; Steppler, *The Common Soldier in the Era of George III*, p.227.
8 Willerd R. Fann, 'On the Infantryman's Age in Eighteenth Century Prussia', *Military Affairs*, vol.41. no.4. (December 1977), p.168; Steppler, 'The Common Soldier in the Era of George III', p.229.
9 Hagen, *Ordinary Prussians*, pp.44–60.
10 Peter Laslett, *The World We have Lost: Further Explored* (London, Routledge, 2004), pp.52–58.
11 Geheimes Staatsarchiv Preußischer Kulturbesitz, (GStaPK) Hauptabteilung (HA): X, Rep. 37 Stavenow, Nr.688.
12 G.E. Mingay, *English Landed Society in the Eighteenth Century* (London: Routledge, 1963), pp.233–235.
13 Hagen, *Ordinary Prussians*, p.124; Laslett, *The World We Have Lost*, p.90.
14 Hagen, *Ordinary Prussians*, pp.278–278, Laslett, *The World WeHave Lost*, p.150.
15 Hagen, *Ordinary Prussians*, pp.262–268; Takashi Iida, 'Coping with Poverty in Rural Brandenburg: The Role of Lords and State in the Late Eighteenth Century', in Tanimoto Masayuki and Wong R. Bin (eds), *Public Goods Provision in the Early Modern Economy:*

and depending on their particular status, often had access to livestock.[16] Families worked hard together to avoid hunger and poverty; indeed, Prussians spoke of their communities not in terms of family but using terms such as 'household' and 'economy'.[17] Despite the economic and labour-orientated nature of these terms, soldiers' letters display a great deal of emotional focus which underlines the importance of the family in their mental worlds.

As this chapter will show, many soldiers wrote to their wives, and it may seem natural to assume that most eighteenth-century soldiers were married. This assumption demonstrates one of the challenges of using letters as a primary source base. From the available period muster rolls of the Austrian, British, and Prussian armies, it appears that an average of 21 percent of soldiers were married during the era under examination. The Austrian army was one of the earlier professionalized armies and provides a useful baseline of comparison for the British and Prussian experience. Christopher Duffy's lifetime of work in the Austrian War Archives has indicated that of the 122,435 soldiers listed in the muster rolls from the Seven Years War era, approximately 14 percent were married. Don Hagist has shown that perhaps 17 percent of British soldiers stationed in New York City were listed as having a woman on the establishment of the regiment.[18] The figure being married could be higher, as some soldiers might have left their wives at home in Britain. Finally, in the Prussian army in the period directly after the Seven Years War, Beate Engelen demonstrated that perhaps 29.65 percent of soldiers in the Berlin and Potsdam garrisons were married. Christopher Duffy, examining just the Potsdam garrison, arrives at a higher figure: 32.2 percent.[19]

These figures, though they come from the period after the Seven Years War, roughly match what historians might expect given the surviving letter recipients from the British and Prussian armies. As an army which recruited heavily from its own population at differing stages of life, the Prussian army was more likely to possess married soldiers. Soldiers who were furloughed in their home villages for nine to 10 months of the year were also likely to have lives that looked more like their civilian counterparts than soldiers who remained with the military permanently. In the period after the Seven Years War, Frederick II deliberately encouraged common soldiers to marry. Thanks to the movement of Russian and Austrian armies, Prussia some 500,000 Prussian civilians had been killed or displaced in the Seven Years War.[20] Frederick II possessed no illusions about the state of Prussia's

---

Comparative Perspectives from Japan, China, and Europe (Oakland, California: University of California Press, 2019), p.118; Roy Porter, *English Society in the Eighteenth Century*, (London: Pelican Books, 1982), pp.290-301.

16 Hagen, *Ordinary Prussians*, pp.222–228.
17 Hagen, *Ordinary Prussians*, pp.222–228.
18 Don Hagist, 'The Women of the British Army in America,' *The Brigade Dispatch: The Journal of the Brigade of the American Revolution*, vols.XXIV–XXV.
19 Beate Engelen, *Soldatenfrauen in Preußen: Eine Strukturanalyse der Garnisonsgesellschaft im späten 17. und im 18. Jahrhundert* (Münster: Lit, 2005), pp.88–89; Duffy, *The Army of Frederick the Great*, 2nd Edition (Chicago: Emperor's Press, 1994), p.81.
20 Matt Schumann, 'The end of the Seven Years' War in Europe,' in Mark Danely and Patrick Speelman, *The Seven Years War: Global Views* (Boston: Brill, 2012), p.514.

economy. Taking drastic and severe measures, he forcibly abducted teenagers (both boys and girls) from neighbouring states under Prussian control (Saxony and portions of Poland).[21] The boys were placed into the army, and the girls were married to Prussian soldiers. This act was understandably later viewed with some embarrassment in Prussia, even if contemporary foreign observers did not find it shocking. Frederick was willing to take whatever steps necessary to rebuild Prussia, regardless of the lives affected. In fact, Frederick's political testament of 1768 positively encouraged marriage and procreation as a solution to demographic challenges. Frederick ordered his officers to 'grant free permission for all cantonists and native soldiers to marry when they request it. This will populate the country and preserve the stock, which is admirable.'[22] As the letter data demonstrates, marriages grounded men in the world of the civilian population.

While Frederick II firmly agreed with the practice of marriage for soldiers, he preferred his officers remain perpetual bachelors.[23] Military intellectuals debated this policy in the British Army as well. During the Seven Years War, British Army chaplain William Agar published a series of sermons in 1758, calling for the total number of recognized wives per battalion to be increased from 60 to 200 (approximately one-third of the paper strength) and defending the virtues of married soldiers.[24] For Agar, having more married soldiers would provide a better family support network in the army, and Agar also believed that being married would cause soldiers to fight more effectively.[25] Although both Frederick II and Agar believed that more soldiers should be married, the simple fact is that married soldiers were a minority in eighteenth-century armies. Our subsequent investigation of letters will show that married soldiers had more of an incentive to write home.

Prussian Soldiers' Letters

These soldiers found military life arduous. Despite having pride in their service, Prussian soldiers were not afraid to admit that it was difficult and demanding. After recalling his glorious first night in the Prussian army, where he was feasted by his new comrades, enlistee Joseph Ferdinand Dreyer sombrely noted, 'after that, I often had to make do with issued bread and well-water.'[26] Prussians soldiers were subject to harsh military discipline and difficult conditions of service. The Prussian military justice system was designed to enforce officers' control over their men and dissuade soldiers from desertion. In addition, soldiers faced a rigorous and physi-

---

21 Franz A.J. Szabo, *The Seven Years War in Europe, 1756–1763* (London: Pearson Longman, 2008), p.423.
22 Frederick II and Gustav Berthold Volz, *Die Politischen Testamente Friedrich's des Großen* (Berlin: Hobbing, 1920), p.83.
23 Duffy, *The Army of Frederick the Great* (1996), p.81.
24 William Agar, *Military Devotion: or the soldier's duty to God* (London: Brindley, 1758), p.xxix.
25 Agar, *Military Devotion*, pp.xxix–xxx.
26 Jospeh Ferdinand Dreyer, *Leben Und Taten eines Preussischen Regiments-Tambours von ihm selbst beschrieben in seinem 93ten Lebensjahre, Altpreussischer Komimiss 22* (Osnabrück: Biblio, 1975 [1810]), p.14.

cally demanding time while on military service. At the height of the Seven Years War, Prussian generals such as Prince Henry, Frederick II's brother, drove men to rare feats of endurance, such as a 148-kilometer march over a period of 72 hours. While not uncommon in the modern era, these gruelling marches were unique in the eighteenth century.

These hardships were easier to bear when soldiers could draw on village connections, both in and out of their regiments. As a result of serving away from their local communities, Prussian soldiers formed firm emotional bonds with both place and family in the eighteenth century. Soldiers expressed emotion about leaving home. When embarking for the onset of the Seven Years War, veteran soldiers wept upon departing from their families: 'There were streams of tears on all sides, and many a veteran hussar rubbed his eyes with the back of his hand, after noticing his wife and children in the circle of saddened spectators.'[27] A private in the Itzenplitz Regiment recalled, 'Now the drums beat the march, and there were streams of tears from citizens, soldiers wives, whores, etc. Likewise the native soldiers of the country who were leaving their wives and children behind, were cast down, full of grief and sorrow.'[28] Returning from abroad with a recruiting party in 1755, a potential recruit for the Prussian army recalled that upon crossing the Elbe, 'the sergeants expressed great joy, because we now walked on the soil of Brandenburg.'[29] Heavily connected to the civilian world, soldiers experienced separation from their homes and families, which caused them pain as they left to join the royal army.

Of these local connections, the most important were undoubtedly familial ties. Of the 78 known surviving letters from eighteenth-century Prussian soldiers, over 70 of them are addressed to family.[30] The letter greetings usually address primarily one person, and then include a number of other recipients. The data collected from these letters are reproduced in Figures 1 and 2.

The most surprising, and misleading, aspect of this data is the high proportion of letters to extended family. If the correspondence of Christian Friedrich Zander with his extended family in the village of Nitzan is disregarded, the proportion of letters addressed primarily to more distant relatives drops to a similar number as letters

---

27 J.G. Lojewsky, *Selbstbiographie Des Husaren-Obersten Von... Ly; Oder, Meine Militairische Laufbahn Im Dienste Friedrichs Des Einzigen* (Leipzig: Kollmann, 1843), p.9.
28 Ulrich Bräker and Johann Heinrich Füssli, *Lebensgeschichte und Natürliche Abentheuer eines armen Mannes von Tockenburg* (Zurich: Füssli, 1789), p.138.
29 Bräker, *Lebensgeschichte*, p.113.
30 These letters are drawn from the following sources: GStAPK, HA X, Rep. 37, Stavenow, Nr. 496; HHStAW, Fonds 133, No. 11670; HStAD, G 28, F 2017; Johann Jakob Dominicus, Kerler (eds.), *Aus dem Siebenjährigen Krieg. Tagebuch des preußischen Musketiers Dominicus* (Munich, C.B. Beck, 1891) pp.61–66; Georg Liebe *Preußische Soldatenbriefe aus dem Gebiet der Provinz Sachsen im 18. Jahrhundert* (Halle/Salle: Gebauer Schwetschke, 1912) passim; Christian F. Zander, *Fundstücke – Dokumente Und Briefe Einer Preußischen Bauernfamilie: (1747–1953)* (Hamburg: Kovacĭ, 2015), pp.21–82; Curt Jany, 'Briefe Preussischer Soldaten aus den Feldzügen 1756 und 1757 und über die Schlachten bei Lobositz und Prag,' in *Urkundliche Beiträge und Forschungen zur Geschichte des preussischen Heeres* (Berlin: E.S. Mittler, 1901), pp.1–59; Rolf Dieter Kohl, 'Ein Brief des Wiblingwerder Bauernshones Johann Hermann Dresel aus dem siebenjährigen Krieg,' *Die Märker*, vol.28, no.3 (1979), pp.82–84.

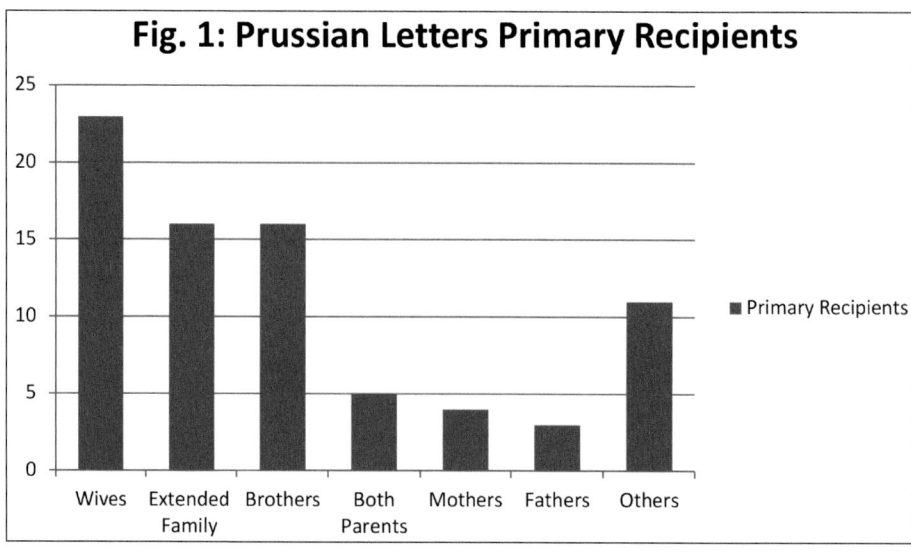

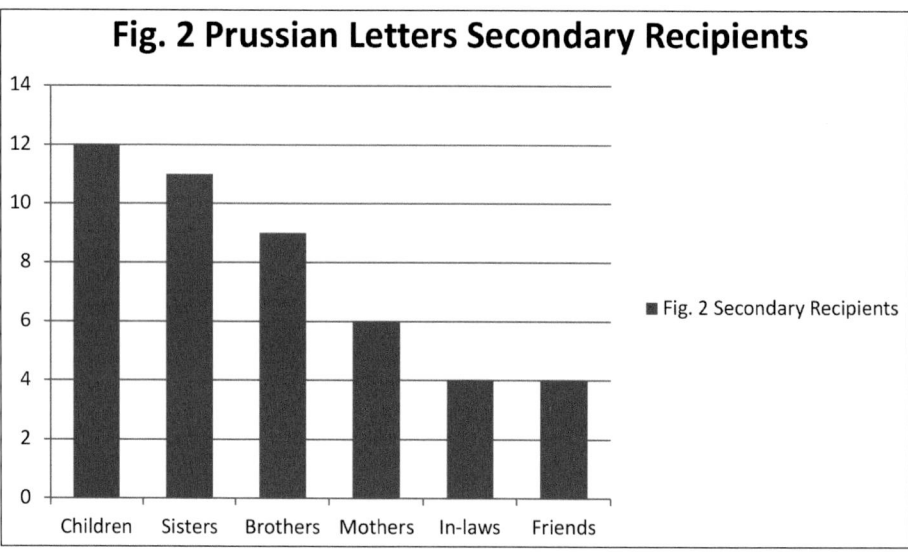

to parents. The data, however, makes one issue abundantly clear: Prussian soldiers whose letters survive corresponded with their wives more than any other group, by a wide margin. Although soldiers' wives managed economic affairs in their absence, pecuniary details fill a rather small proportion of these letters. Rather, the letters are often filled with tender anecdotes, and are signed with sentiments such as, 'I remain your faithful husband.'[31] These greetings clearly indicate that some Prussian cantonists possessed emotional lives, and that they remained connected to their family and local identities. The small sample size of the letters should prevent historians from

---

31 Hans Bleckwenn, *Preussische Soldatenbriefe* (Osnabrück: Biblio Verlag, 1982), p.15.

drawing sweeping conclusions. However, the relative similarity of surviving letters, the frequently included messages from non-literate soldiers they contain, and the existence of a widespread Prussian army post service during the Seven Years War indicates that these could be broadly shared sentiments.

Children, sisters, and mothers were frequently mentioned in letter address lines, but rarely as the primary recipient, possibly as a result of their status as dependents. Brothers and mothers can be found in both primary and secondary recipient lists, both of which might reflect their ability to transcend the dependent position. Mothers could become important figures when fathers died, and brothers would merit a letter addressed directly to them. When mothers were listed in a secondary status, they were almost always subordinate to either wives or fathers. Finally, in-laws were occasionally mentioned but remain the smallest group of family relations present in Prussian soldiers' letters. A few soldiers also greeted non-relations in their address lines, always in a secondary status, and both friends and acquaintances are listed, indicating that these letters, far from being private, took on village-wide importance.

For eighteenth-century soldiers, letters home became communal documents, designed to connect many soldiers both to individual families and to their home villages. The first way in which these letters are communal is literal: Prussian soldiers' letters often contain more than one voice. Messages from other authors frequently appear as postscript or notation in the margins of eighteenth-century soldiers' letters. Sometimes, these are messages written by the author of the main letter; other times, they appear as new voices written in a different hand. In his letter home on 24 November 1756, Christian Friedrich Zander added a post-script: 'P.S. Jochen Müller, Weidemann, Andreas Backe, and Börnicke are all still healthy and send many greetings home. Valtin Bährend is with the Regimental Surgeon, and the entire army says hello. Please write to us again! Farewell.'[32] Christian Zander's greeting, 'the entire army says hello,' does much to capture the spirit of these letters home. In the same letter, written by Christian Zander to his family in Nitzahn, a note in the margins informs the reader, 'We also greet old H. Wilbergen, and if he is still healthy, we can't wait to see him in the flesh when we return home.'[33] Rather than private documents designed for a single reader, these letters became lifelines between groups of soldiers from the same village and their local communities. Often these marginal notes contain pleas for return letters, some as simple as 'write us soon!' Both Johann and Christian Zander scribble marginal notes complaining about the length of time since they received a letter from home.[34]

Historians can observe the communal nature of Prussian soldiers' letters, even messages dealing with matters that might seem private. Hans Wölcke, a farmer's son from the village of Stavenow in Brandenburg, wrote home in 1757 about a rather private issue: he was asking for his former stepmother's hand in marriage. Wölcke and the object of his affections, Liese Hintzen, were around the same age, and had flirted as teenagers, before Hintzen's father had married her off to Wölcke's

---

32 Zander, *Fundstücke*, p.33.
33 Zander, *Fundstücke*, p.33.
34 Zander, *Fundstücke*, pp.41, 64.

own father. After the death of Hintzen's much older husband, she and Hans had continued their relationship, and indeed, had two children together. It was a strange and difficult situation, in fact, both Hans and Liese were later charged with incest.[35] Despite the situation, there was nothing private about Hans's letter home. Indeed, eight other soldiers wrote postscripts or marginal notes home to their loved ones in the same letter as Hans's marriage proposal. 'Niklaus Hintze... greets his wife many thousand times.' Mathias Hintze complained about his wife's lack of correspondence, saying, 'he doesn't know what he should think that she hasn't once written to him.'[36] Even letters that contained sensitive communication between husband and wife still carried messages from other soldiers to the village community.

Even more frequently than their postscripts and marginal notations, the Zanders would include messages from family and friends with the army in the main bodies of their letters. More than just messages from soldiers with the army, these include pleas for letters from home for other men: 'Every day, we wake up hoping to get a letter from you, but it does not come. All of our comrades alike ask for letters from home. Please write more about how business is going, and whether or not the work is finished.'[37] Soldiers' letters became a vehicle for connecting not only the primary author to his family, but for emotionally connecting the entire village, even though its members were physically dispersed. Perhaps even more importantly, communal letter writing kept men from the same village in close contact with one another in the army. In addition to aiding the transit of letters via a free military postal system, Frederick II celebrated soldiers' loyalty to family and community as a motivation for soldiers in difficult times and wrote about it.[38]

This was the emotional core of the sentiments produced by appeals to kin and home made by Frederick the Great in the course of the Seven Years War. In Frederick's famous 'Parchwitz Address,' given on the eve of the Battle of Leuthen in 1757, he appealed directly to family connections: 'Bear in mind, gentlemen that we shall be fighting for our glory, the preservation of our homes, and for our wives and children. Those who think as I do can rest assured, that if they are killed, I will look after their families.'[39] The geopolitics of the Seven Years War made it possible for Frederick to characterize this war as a defensive conflict for the preservation of Prussia, and this narrative resonated with common soldiers, even if, in all likelihood, the king referred only to the families of his officers.

Rudolph Kaltenborn, an old officer writing in the 1790s, recalled that the speech was often repeated by Prussian soldiers 'who could never hear it without tears, and although they were under arms, these rough and unfeeling soldiers would cry like children.'[40] Frederick II of Prussia, in appealing to family and home, understood that he touched his soldiers at a deep emotional level. Frederick II realized that soldiers

---

35 William W. Hagen, *Ordinary Prussians*, p.508.
36 GStAPK, HA X: Rep. 37, Stavenow, Nr. 496.
37 Zander, *Fundstücke*, p.38.
38 Christopher Duffy, *The Military Experience in the Age of Reason*, p.128.
39 O. Herrmann, 'Prinz Ferdinand von Preussen über den Feldzug Vom Jahre 1757,' *Forschungen*, vol.XXXI (1918), pp.101–102.
40 Rudolph Wilhelm von Kaltenborn, *Briefe eines alten preussischen Officiers verschiedene characterzüge Friedrichs des Einzigen Betreffend* (Potsdam: Hohenzollern, 1790), p.53.

fought as a result of cohesive bonds of loyalty and devotion. Much has been made of Frederick's famous statement that soldiers should fear their officers more than the enemy.[41] Indeed, Sascha Möbius has written an excellent monograph showing the tactical reasons this statement could not be true, exploring both the tactical underpinning of the Prussian army, as well as the religious, patriotic, and material motivations of Prussian soldiers.[42] Frederick, in less well-known statements, emphasized the communal bonds of the village, region, and canton-district as the most important element in fostering courage among soldiers. Of his cantonists, he recalled in 1768, 'These... citizen-soldiers [soldats-citoyens] are all from the same locale. Many of them are relatives and know one another ... these cantons give encouragement and bravery, for friends and relatives who fight together, do not give up easily.'[43] In the case of native cantonists, local connections to friends and family provided more cohesion in combat than the threats of officers, as demonstrated by the heavy casualties sustained by units in combat, such as Regiment von Itzenplitz at Hochkirch and Frederick II's half of the army at Torgau.

Military observers and authors, such as the comte de Mirabeau and the officer Jakob Mauvillion observed the Prussian army after the Seven Years War. These writers asserted that local connections were at the heart of Prussian military success. In his view, soldiers' 'connections to their homes were not just beneficial, they guaranteed the victories of Frederick II.' He also found that 'the men of each regiment are drawn from the same province, and all are well acquainted. This makes their lives easier and creates a sort of comradeship most useful on the day of battle.' Like Frederick, Mirabeau thought that local connections were the glue that held the Prussian army together under fire. The French nobleman even gave some prescriptive advice: why not name the regiments after their canton districts, rather than their commander? 'The embodied spirit could be perfected still further, if the regiments were named after their districts, rather than their commanders.'[44] As revealed by their letters, the writings and speeches of their king, and foreign observers, the Prussian army contained a high degree of loyalty to place and family. This localism translated well into regimental loyalty. The Prussian regiments, however, were not the only eighteenth-century force which embodied this localism, and the chapter will now turn to evidence of local loyalty in another army.

Jacobite Soldiers' Letters

Letters from Jacobite soldiers to their families indicate local attachments resembling those found in Prussian soldiers' correspondence. This portion of the

---

41   Frederick II, *Instruktion für die Commandeurs die Infanterie-Regimenter*, 11 May 1763.
42   Sascha Möbius, *Mehr Angst vor dem Offizier als vor dem Feind?: Eine mentalitätsgeschichtliche Studie zur preußischen Taktik im siebenjährigen Krieg* (Saarbrücken: VDM Verlag, 2007), pp.16–28.
43   Friedrich II and Volz, *Die Politischen Testamente Friedrichs des Großen*, p.139.
44   H. G. Mirabeau and J. Mauvillion, *Systeme Militaire de la Prusse et Principes de la Tactique Actuelle des Troupes les plus Perfectionnees. Extrait de la Monarchie Prussienne* (London, 1788), p.78.

chapter examines the letters of Scottish Jacobite soldiers during the 1745 Jacobite Rebellion. During the uprising, a part of the larger War of Austrian Succession, Prince Charles Edward Stuart led an attempt to overthrow the ruling Hanoverian dynasty of Great Britain and restore the Stuart family to the throne. After landing in Scotland with a few supporters, the Prince received extensive support from the French, raised a significant army, and won victories in Scotland. Following these successes, the Jacobite force invaded as far south as Derby in England before eventually being driven back and defeated at the Battle of Culloden. There is a great debate about the exact nature of the 'Highland Army' which the Stuart Prince led, and Stuart Reid argues that essentially it consisted of four types of soldiers: first, highland troops who initially served as a result of social obligation to their lords; second, men who were selected for service as militiamen; third, troops who were forcibly conscripted by threat of force; and finally, men who willingly volunteered for service.[45] Like much of the scholarship which challenges the traditional romantic view of the Scottish soldiers of this rebellion, Reid, together with Christopher Duffy and Murray Pittock, point to the idea that many Jacobite soldiers were conventionally equipped like most European soldiers and organized into relatively advanced operational units, but they also maintained a connection to their traditional way of fighting.[46] Far from being a wholly traditional force, the Jacobites (like the Prussian army) seem to have relied on localism and kinship ties even as they partially adapted to new ways of fighting. Even as Jacobites were armed with new French muskets, and drilled in more modern combat methods, their army remained organized by clan regiments, which drew on kinship ties. A set of correspondence captured by the Hanoverian government in late 1745 has survived, offering a snapshot of Jacobite soldiers' letters. Although most of the 60 letters come from officers, perhaps 20 come from unidentified or low-ranked men. The recipients of these letters are indicated in Figure 3.[47]

Like Prussian soldiers, Jacobites primarily wrote home to their wives. The only explicitly identified common soldier among the letter authors, Sergeant Duncan Macgillis of Macdonell, of Glengarry's Regiment, wrote to his fiancée Margaret on 30 October 1745. Margaret had previously sent Duncan a letter, and he acknowledged the receipt of it, saying he was 'glad to hear that [she] was in good health.' Furthermore, Duncan pointed to the network of local connections supporting the Jacobite army: Margaret had previously sent a letter by way of the son of a man known to both of them, Allan Roy, but this letter, for some reason, had not reached the army. He indicated that he missed Margaret: 'we are in opinion every day to march on to England and being a sergeant and having the trouble of the company, and God knows how soon I can present my love to you, and nevertheless my love

---

45 Stuart Reid, *1745: A Military History of the Last Jacobite Uprising* (Staplehurst: Spellmount, 1996), pp.199–201.
46 Reid, *1745*, pp.204–205, Christopher Duffy, *Fight for a Throne* (Warwick: Helion, 2015), pp.312–339; Murray Pittock, *The Myth of the Jacobite Clans* (Edinburgh: Edinburgh University Press, 2009), pp.163–182.
47 All letters are drawn from TNA, SP 54/26/122, Letters from Jacobite soldiers to family and friends..

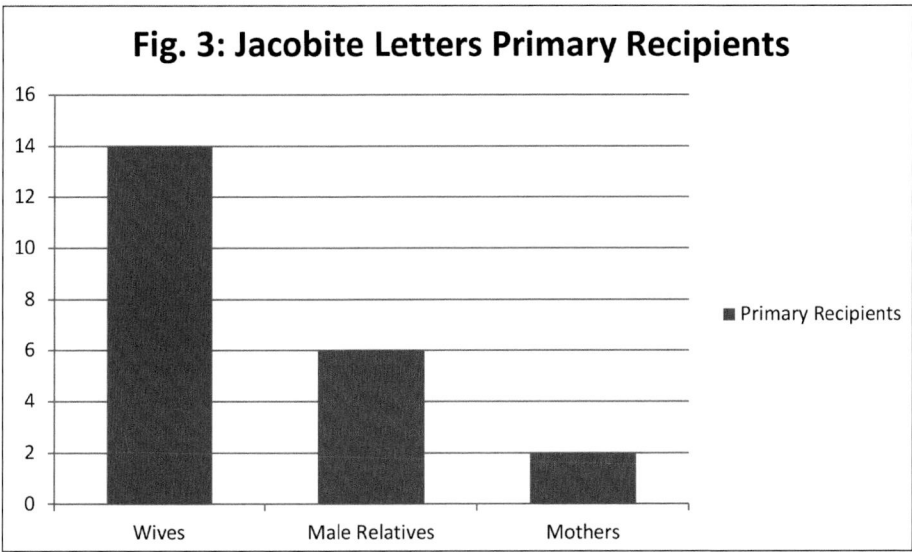

Fig. 3: Jacobite Letters Primary Recipients

is as constant to you as it was formerly.' A close reading of the letter indicates that Duncan and Margaret were not yet married, as he both indicates that they will 'make all things complete' upon his return to Scotland, and furthermore writes that she should 'give my service to your children,' indicating that these could be children from a previous marriage. Duncan's single letter gives a sense that literate common soldiers in the Jacobite army were capable of maintaining a correspondence with their loved ones via local connections during the war of 1745–1746.[48]

Another letter, from Donald Macdonell, likely a common soldier in the Duke of Perth's Regiment, refers to previous letters carried by a local boy, to whom Donald refers as 'young Leek.' Like Duncan Macgillis, Donald writes to his mother to inform her of the possibility of an invasion of England, and that he might be away for some time. He transmitted some of his pay home for his mother with the letter, and also asked the same 'young Leek' to give some ribbon for his sister as a present. Macdonell's letter does not show any awareness of the overall Jacobite operational plan, and he mentions with surprise that other regiments were joining the portion of the army he was stationed with, notably the McPherson Regiment and 'Athole People.' In the absence of a formalized military post system, using acquaintances to carry letters drew on the Jacobites' local ties, and attempted to keep the army and its constituents connected.[49]

Of this collection of letters from Jacobite soldiers, one letter stands out. On a single sheet of paper, there are two distinct messages, written in slightly different hands, likely from two brothers or close male relatives, Evander and Alexande McIver. Evander wrote to his mother, Rorie McIver, seemingly regarding a dispute with their local gentry. He noted, 'I have written to the lard [sic] about what you requested me in your letter and have delivered to his honer to send me word how he

---

48   TNA: SP 54/26/122, f.363.
49   TNA SP 54/26/122, f.367.

has a mind to do with you.' Evander continued, noting that the laird was obligated to him as a result of his military service, and that his mother should 'send me an answer of this with the first opportunity and send me world how my wife is and all friends at home[.]' On the same sheet of paper, Alexande McIver wrote to his wife directly, assuring her of his 'kind love' and asking that she write to him via his brothers. Once again, local and family connections facilitated the carrying of letters across the British Isles. In a marginal note, Alexande begged his wife to 'give my kind services to my mother-in-law and my brothers.' There is also a slight indication that Alexande felt somewhat shy about writing intimate details to his wife in a letter which was a joint writing from Evander: 'I would be kinder but not having the opportunity, I hope you will excuse me.' This particular letter demonstrates the local and familial nature of military letter-writing between Jacobite common soldiers and their families during the last great Jacobite rising. Although occasionally noting military details, these men were primarily concerned with maintaining a connection to their wives and loved ones, using their family and local connections in order to transmit their letters.[50]

Both Prussian and Jacobite soldiers wrote to similar correspondent groups. Though both their armies and families were structured in slightly different ways, Prussian and Jacobite soldiers both drew on local connections in their letter writing. The surviving letters from Prussian and Jacobite soldiers draw on similar motifs: assurances of love, a desire to be reunited with the family unit, and requests to be remembered to extended family and friends. There were differences, though. For example, Hans Wölcke's inclusion of details regarding an ongoing incest trial in a letter with messages for other soldiers contrasts strongly with Alexande McIver's shyness regarding affection to his wife in a communal letter. This could be a result of personality or cultural differences. On the whole, however, Prussian and Jacobite letter-writing cultures share many similarities. Some of these similarities are shared by all soldiers attempting to communicate with their loved ones, but the local nature of military recruitment in both portions of the Prussian and Jacobite militaries also help to explain some of these common threads. Having examined two comparable cultures of letter writing among eighteenth-century soldiers, the remainder of the chapter now examines a third which shows significant differences: the British Army which served across the globe between 1733 and 1783.

British Soldiers' Letters

Both before and after the 1782 Army Reforms, British soldiers were sometimes, though not always, disconnected from the locality from which they were drawn. On multiple occasions young British men used military service in order to escape what they viewed as domestic problems, such as fatherly responsibilities or an unfavourable apprenticeship.[51] British letter writers demonstrate that this hasty departure

---

50   TNA SP 54/26/122, f.375.
51   For examples of this type of soldier, see the memoirs and letters of Sampson Staniforth, Duncan Wright, and Samuel Hickson.

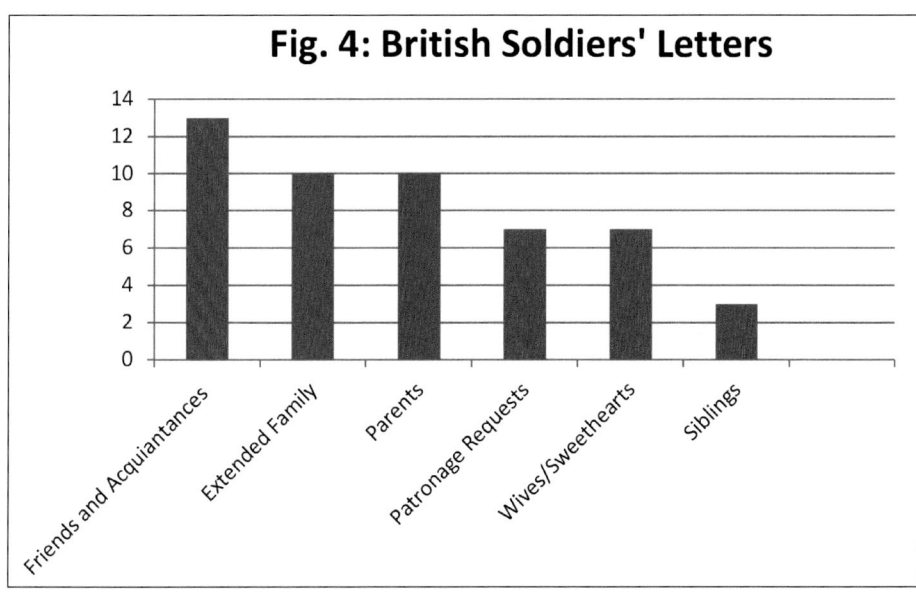

could cause a rift with their local communities, though not all letters reflect soldiers who have abandoned their responsibilities or created tension in their communities. Those who did leave under dubious circumstances, however, often used their letters as a way of asking for forgiveness for abandoning their domestic responsibilities in joining the army. With this said, there are definitely examples of soldiers whose correspondence shows a continued devotion to their kinship networks and local communities, much in the same way that Prussian and Jacobite soldiers did. Approximately 60 letters are included in the dataset which makes up Figure 4.[52]

Unlike the Jacobite and Prussians letter writers, these British soldiers did not write most frequently to their wives; indeed, letters to wives and sweethearts make up a very small portion of the total surviving correspondence of British soldiers. Rather, the nature of their enlistment frequently meant that they were unmarried and unattached at the time of recruitment. As a result, although they frequently wrote to their family – both their parents and extended family – they also wrote home to friends and acquaintances. This lack of correspondence between soldiers and their wives

---

52 These letters are drawn from: TNA, FO 95/5/3; HCA 30/272/3; HO 42/46/32; SP 36/72/124; SP 36/84/2/8; WO 28/8/137; The National Army Museum (NAM), 1976-07-40; 1986-11-1; 2008-06-4; BL, Letters and other Papers of Samuel Hickson; Bedfordshire and Luton Archives and Record Service: R 769; Gloucestershire Archives (GA), D4582 Bowly Family of Cirencester; D153 Jackson Family of Sneyd Park; The Berkshire Record Office, R/D/134/13; Lancashire Archives, DDX 2743/MS5237; QSP/1996/9; Dorset History Centre, D/WIB/C/93; D/HAB/F17; Lambeth Palace Library, Beloe Papers, MS 3263, f. 148; National Records of Scotland (NRS), GD248/509/3/74; William L Clements Library, Schoff Revolutionary War Collection; Society of the Cincinnati Library, MSS L1992.1.477; John Wesley, Thomas Jackson (ed.), *The Journal of the Rev. John Wesley, A.M.*, (London: Wesleyan Conference Office, 1869), vols.1 and 2, Anon., *British Glory Reviv'd* (London, J. Roberts, 1743); E. Linn, 'The Battle of Culloden', *Journal of the Society for Army Historical Research*, (1921–1922).

does not necessarily mean that soldiers in the British service esteemed their families or local communities any less than Prussians or Jacobites, but with a lack of a clear correspondent (a wife) who literally represented them in their village community, they were forced to seek out intermediaries who could represent them to a family who might have been angered by their departure.

For soldiers who fled from their family responsibilities, writing a letter was a vital first step for the eventual reintroduction to the village community. A first letter home, not directed to the immediate family but to a trusted third party or friend, could do much to heal the initial emotional wound of flight or anger which had separated the soldier from his village community in the first place. Samuel Hickson, who fled from his family to serve in the British forces on the subcontinent of India, wrote two letters to friends and acquaintances before writing to his mother. First, he wrote to a prominent man of business in his home area, Mr Egeer, in order to ask for money and patronage. He followed that letter, almost immediately in the summer of 1779, with a letter to his former schoolmaster, Mr John Smith, begging him to be an intermediary between Hickson and his family. Concluding the letter, Hickson asked for a final time, 'let my mother, relations and friends… know that I am alive.' Hickson only followed up to a member of his immediate family later, writing his mother in December of 1781: 'I would have you free from all apprehension as to my situation.'[53] Hickson realized that his flight from the family farm had placed his mother in a difficult situation, and continued his letter:

> I shall eagerly embrace the first opportunity to return to you… I sincerely hope that God will raise up friends to protect you from distress and that my return, if it should please God to grant me that indulgence, I may find you and my brothers and sisters happily, give my kind love to them all and believe me to be your truly affectionate though unhappy son.[54]

In this passage, Hickson implores God to pick up his own economic slack, and also seems serious in his desire to return home. By leaving the family for India, Hickson placed his mother in a position where she would need economic assistance. Conscious of this embarrassment, Hickson used an intermediary to broker relations with his family, and then employed religious language in order to demonstrate his sincerity in desiring to help his mother economically and be reunited with his family.

Not content simply to ask God for assistance in assisting his family, Samuel Hickson wrote to one of his male relatives, his cousin Mr J. Hickson, in order to express his gratitude for the cousin's help in caring for the mother and siblings the soldier had left behind. Samuel Hickson had received a letter from his brother William, informing him,

---

53 BL: European Manuscripts, #B296, Letters and other Papers of Samuel Hickson, Letter of 24 December 1781.
54 BL: European Manuscripts, #B296, Letters and other Papers of Samuel Hickson, Letter of 24 December 1781.

that you [J. Hickson] live in the house in which I left my mother, and that you are very kind to her. This has given me a great deal of satisfaction as I have suffered great uneasiness from the fear that she might be in distress in her old age, without a friend, whilst I was at such a distance that it was absolutely impossible I should assist her.[55]

Samuel, in writing to his cousin, expresses both gratitude and remorse, simultaneously downplaying the nature of his absence. Hickson continued to his cousin, 'I hope and trust you will continue your kindness to her for my sake, as well as the whole family, and should providence ever put it in my power to make you a return for it, it will be a great pleasure to me.' Here, Hickson continues his gratitude and follows it up with a vague indication of financial reward, once again couched in religious language: 'All I can do at present is to return you my sincere thanks and give you an account of the transactions in my life since I left you. By the perusal of which if your curiosity is in any way gratified, it is the only demonstration of my gratitude you can at present receive.' Here, Hickson offers the story of his military service as cultural currency in lieu of actual financial reward for his cousin's care for his immediate family. Hickson demonstrates the strength of kinship ties and local communities; indeed, his flight into the army forces these networks to care for dependent members of their communities. Hickson's letters provide a window into the sheepishness which soldiers felt after abandoning their social responsibility to their families, and the religious and cultural steps they took to excuse their behaviour and find embrace in their village communities once again.[56]

Other British soldiers entered the army with the blessing of their village communities and maintained a close correspondence with them throughout their military careers. A cavalry soldier named Hooper wrote to his wife just after the Battle of Dettingen in 1743. Most of the letter is a fairly mundane description of the battle. On the same sheet of paper as this letter, however, John Griffith, a drummer in the same regiment as Hooper, wrote a short inscription to his mother and sister, evidence of strong family and local connections in the British Army in the same manner as the Prussian army.[57] Another example of this type of familial mail is a letter from Thomas Plumb, a soldier in Captain McDonald's company of the 22nd Regiment of Foot, who wrote home to his family on February 22, 1777, while serving in North America. Plumb addressed his letter primarily to an Alexander Johns, to whom he referred as his brother, and was likely his brother-in-law. The letter indicates that Plumb's wife was in close contact with Johns; the secondary recipients includes 'wife, child, and all enquiring friends'. Plumb wrote frankly regarding the challenges of combat during the war, and indicated potential problems facing the British force: 'our present state and situation in this country at the present time our duty is very hard'. Plumb, like the Prussian and Jacobite soldiers encountered above, greeted a host of relations near the close of his letter: 'my kind respects to my

---

55 BL: European Manuscripts, #B296, Letters and other Papers of Samuel Hickson, Letter of 12 June 1782.
56 BL: European Manuscripts, #B296, Letters and other Papers of Samuel Hickson.
57 Anon., *British Glory Reviv'd* (London: J. Roberts, 1743) pp.42–43.

loving wife and child, Uncle Wood, Molly, and Little William, and all enquiring friends.' The letter provides a clear indication that for some British soldiers, family, kinship, and local communities remained a constant source of interest and encouragement as they were deployed on other continents. As he was finishing the letter, a final thought struck Plumb. Out of space in the body of the page, he moved to the margins and wrote a final understated but firm instruction: 'Your [ans]wer by the first opportunity.'[58]

Like Plumb, soldier William Calder was worried that it had been a long time since he had received a letter from his extended family: 'Let me know... how many letters you got from me since I left you, this is the fourth and I should like to know whether you get them or not.'[59] Like the Jacobite soldiers earlier in the chapter, Calder used friends to carry letters to his family. His letter dated 10 February 1781 was carried to his family in Glasgow by a sergeant on recruiting service there, Sergeant Boyle. Calder, who informed that Boyle 'would be glad to drink a bottle of ale or Dram with you on my account or with any of my relations.'[60] Sergeant Boyle also carried Calder's letter of 3 June 1782, which informed the family of the discharge of their kinsman from the military. Corporal Brock carried Calder's letter of 25 May 1784. These letters clearly point to the absence of a reliable military post system, but also indicate that soldiers who were determined to correspond with their families found ways around these obstacles. British soldiers, then, wrote letters that simultaneously match and deviate from the patterns found in the Prussian and Jacobite letters. Some British soldiers, like their Prussian and Jacobite counterparts, wrote to their families and village communities, clearly displaying emotional bonds to both. Others, having fled from their social responsibilities into the army, attempted to use letters as a way of mediation in their village worlds following their departure.

The Village World and Negotiated Authority

As the next chapter will show, European military observers became convinced that in Prussia harsh discipline had produced a superior sort of soldier, who would unthinkingly follow the tactical directives of his officers. In many ways, this is the origin of the idea that eighteenth-century soldiers were robotic automata. In the twentieth century, this led Otto Büsch to argue that social-militarization had occurred in this period: the harsh treatment of robotic soldiers, in his view, allowed for landlords to treat their peasants with equal brutality. He argued, 'The *Junker* [Prussian officer/landlord] on the exercise field could accustom himself to the unlimited practice of punishments of which the rural subject back on the

---

58   TNA: HCA 30/272/3, Box 82J, Thomas Plumb to Brother, 22 February 1777.
59   NAM: 1986-11-1-1. Nine manuscript letters written by William Calder, Sergeant 3rd Guards, to his aunt and uncle at Paisley 1778-1785.
60   NAM: 1986-11-1-5. Nine manuscript letters written by William Calder, Sergeant 3rd Guards, to his aunt and uncle at Paisley 1778-1785.

estate became the object.'⁶¹ In this view, the brutalized robotic soldiers infected their society with subservience.

Perhaps unsurprisingly, scholarship on German social development has taken many twists and turns since Büsch penned this argument in the 1960s. Since the 1980s, a new wave of scholarship overturned this model. This generation of scholarship on the Prussian army and society is best exemplified by writings from Martin Winter, who asserts that:

> Neither the observed behaviour of those affected by cantonal service, nor that of the mid-level authorities allows for the conclusion that during the eighteenth century the Prussian military system created a particularly submissive people on the far side of the Elbe. It is to be hoped that more research into Brandenburg-Prussia, which deals with the actual living and service conditions of soldiers, will paint a more differentiated picture of the network of relationships between the military and society.⁶²

Instead, both peasants as soldiers and civilians felt comfortable approaching appointed authorities if they believed that something unjust was occurring. Importantly, Katrin and Sascha Möbius have demonstrated this in two cases involving soldiers from the Itzenplitz Regiment (IR 13).⁶³ In the first case, soldier Christian Friedrich Zander attempted to use his rights and village connections to obtain release from military service, demanding a redress of grievances from his superior officers, and through a related intermediary, to both the king himself and *Generalmajor* Phillip Bogeslav von Schwerin. In the second case, Johann Dietrich Zander attempted to prevent an officer from transferring a man from his home village to the grenadier company of the regiment. Though unsuccessful in their petitions, the Zanders demonstrate that a village culture of rights and traditions had been transferred into the Prussian army. The early modern world of negotiated authority, present in the civilian world of the eighteenth century, was maintained in a military setting by the close connections between Prussian soldiers and their village worlds. How was this connection maintained?

The data from letters present shows that married soldiers retained a firm connection to their civilian communities. Soldiers also retained a connection to the civilian world when they actively laboured in civilian trades. In terms of their daily life and work in peacetime, eighteenth-century Prussian soldiers shared much with civilians. Bräker's view of soldiers and civilians working together speaks against the

---

61 Otto Büsch, *Militärsystem und Sozialleben im alten Preußen, 1713–1807: Die Anfänge der sozialen Militarisierung der preußisch-deutschen Gesellschaft* (Ulm: Walter de Gruyter & Co., 1981 [1962]), p.43.
62 Martin Winter, *Untertanengeist durch Militärpflict: Das Preussische Kantonsystem in brandenburgischen Städten im 18. Jahrhundert* (Bielefeld: Regionalgeschichte, 2005), p.466.
63 Katrin and Sascha Möbius, *The Psychology of Honour: Prussian Army Soldiers and the Seven Years War*, (London: Bloomsbury Academic, 2019), pp.197–203. For a fuller discussion of these issues, see: Katrin and Sascha Möbius, 'Fighting in Frederick II's Favorite Musketeer Regiment: A Unique Series of Prussian Soldiers' Letters from the Seven Years War', in Burns (ed.), *The Changing Face of Old Regime Warfare*, pp.228–247.

argument that soldiers and civilians developed mutually exclusive identities. Rather, as Beate Engelen has pointed out, since the era of Frederick William I, civilians and military effectively coexisted in the urban spaces of eighteenth-century Prussia.[64]

Likewise, in rural Prussia, furloughed cantonists worked as farmers for much of the year. William Hagen has argued that these soldiers were occasionally a disruptive and violent force in village life, using examples such as of furloughed soldiers getting drunk and disorderly at weddings, becoming involved in petty crime, and exploding hand grenades as a practical joke.[65] It is important to realize that using official village and court documents may skew the picture somewhat: court records are likely to produce a rather dim view of human nature. Villagers also often resented the special status soldiers were afforded as cottagers (*Büdnerei*). It is also clear, however, that soldiers retained civilian goals, using their military status as a means to advance themselves in civilian life by becoming cottagers, marrying well, and establishing themselves as heads of household. Hagen's sources, however, also reveal a world in which civilians displayed violence towards soldiers: cursing and threatening them with violence at legal hearings, beating them during disputes over game, and threatening to clobber them with rocks.[66] Tellingly, there were furloughed soldiers on both sides of some of these conflicts, so claims that soldiers acted as a group hostile to village interests appear misguided.[67] It is doubtlessly correct that the presence of young and armed men disrupted the village communities with occasional violence. With that said, it is also clear that violence marred the worlds of eighteenth-century villages without the presence of soldiers, who on the whole worked alongside their relations, and attempted to better themselves on their furloughs to the village community.[68]

Foreign observers of Prussian life also commented on the working lives of soldiers. Jacques Antoine, comte de Guibert, travelled to eighteenth-century Prussia shortly after the Seven Years War. There he observed that off-duty soldiers roamed freely 'without uniform of any kind, dirty, uncombed, ragged, going about just as they pleased. There are soldiers on every street corner, pursuing every means of employment imaginable.'[69] Ilya Berkovich has observed that the fact that Guibert could tell the men were soldiers was a sign that military life separated them from the civilian population, but the ragged, unkempt appearance of the soldiers undermines that interpretation. Guibert clearly depicts soldiers mixing freely with the civilian population, harbouring no hostile attitudes toward them. Guibert concludes his remarks with an observation that would ring increasingly true in both military

---

64 Beate Engelen, 'Fremde in Der Stadt: Die Garnisongesellschaft Prenzlaus Im 18. Jahrhundert,' in Jürgen Theil, Olaf Gründel, and Klaus Neitmann (eds), *Die Herkunft Der Brandenburger: Sozial- Und Mentalitätsgeschichtliche Beiträge Zur Bevölkerung Brandenburgs Vom Hohen Mittelalter Bis Zum 20. Jahrhundert* (Potsdam: Berlin-Brandenburg, 2001), pp.116–120.
65 Hagen, *Ordinary Prussians*, pp.466–472, 477.
66 Hagen, *Ordinary Prussians*, pp.466–472.
67 Hagen, *Ordinary Prussians*, p.469.
68 For more on this point, see David Warren Sabean, *Property, Production, and Family in Neckarhausen* (Cambridge: Cambridge University Press, 1991).
69 Jacques-Antoine-Hippolyte, comte de Guibert, *Journal D'un Voyage En Allemagne, Fait En 1773* (Paris: Würtz 1803), vol.1, p.166.

and civilian contexts across Europe in following centuries: 'In Prussia they proceed from the principle that no kind of occupation can demean a soldier, as long as it brings in money.'[70] British observers noted with surprise that the Prussian army was quartered with the civilian population, living in singles and doubles among the houses of Berlin, Königsberg, Küstrin, and Magdeburg. John Moore, a British traveller in Prussia, was shocked by the presence of soldiers' laundry hanging out the windows of civilian homes, and cited a few reasons for quartering troops on the civilian population: 'that a connection and good-will may be cultivated between the soldiers and their fellow-citizens; and that the former may not consider themselves as a distinct body of men, with a separate interest from the rest of the community.'[71] In short, Prussian soldiers occupied a special place within Prussian society, but remained integrated into society, rather than wholly distinct from it. Furthermore, eighteenth-century Prussian soldiers remained an active economic part of civilian society in ways which modern soldiers in the western world do not. 'Interpenetration' between civilian and military society could at times cause friction when soldiers felt abused, but working alongside civilians remained a large part of Prussian soldierly experience.

This chapter asserted that the local village communities formed an important part of soldierly worldviews in Prussia and the British Isles during the eighteenth century. Some soldiers, most of whom hailed from rural communities, maintained contact with those communities via the facility of letter writing. Family bonds and local connections helped construct Frederick II's army into the formidable fighting machine that survived, and therefore won, the Seven Years War. Soldiers stood by one another in the horrors of combat, not for some abstract idea of nationalism, or even a developing notion Prussian patriotism, but because they were sustained by their local communities. These communal ties also supported soldiers as they negotiated authority with their officers. These local communities followed soldiers to the army, in the form of village comrades in the same unit, and communal letters from their villages. These emotional bonds, between soldiers within their regiments, and between soldiers and their village families at home, helped the Prussian army to sustain the horrible losses it suffered in the Seven Years War. Men in other armies, such as Jacobite and British soldiers, reflected similar but not identical connections to village life. The available sources make it clear, however, that the canton system and the military-civilian relationship in Prussia allowed for fuller development of these emotional bonds. These connections, particularly in the Prussian case, are important to keep in mind as we move to the next chapter, which explores the phenomenon of European Prussomania, which misunderstood the Prussian army as a robotic and tightly controlled force.

---

70 Guibert, *Journal D'un Voyage En Allemagne, Fait En 1773*, vol.1, p.166.
71 John Moore, *A View of Society and Manners in France, Switzerland, Germany, and Italy: With Anecdotes Relating to Some Eminent Characters* (Philadelphia: R. Bell, 1783), vol.2, p.117.

4

Theories of Battle and Prussomania: The Origins of the Clockwork Soldier Myth

> 'An Army formed of good Officers moves like Clock work'
> George Washington to John Hancock, 1776[1]

Two Clockwork Tales

In 1733, Dublin printers circulated a humorous article, calling for the disbandment of the British standing army, and its replacement with an army of wax and clockwork. This article told the story of the author's friend, who when received at a court of the *Kleinstaateri* in Germany, noted that the small-holding prince had assembled soldiers for his reception. 'To tell you the plain truth,' the German prince replied, 'my men are of wax, and exercise by clock-work.' The author of the text then humorously advised disbanding the standing army of Great Britain, since 'for these last five and twenty years, our Land Forces have been of no use whatsoever.' Indeed, the author asserted that officers tried in vain 'to bring our present army up the… perfect of a waxen one; it has prov'd impossible to get such numbers of men all of the same height… timing exactly together the several motions of the exercise'. Indeed, the author argued, would not it be better to have machine soldiers, since 'their exercise will… be in the highest German taste, and may possibly arrive at *the one motion* [simultaneity], that great [desire] of our discipline.'[2] This pamphlet combined poking fun at the Hanoverian dynasty with humorous observations on the seeming absurdity of contemporary military culture.

Later in the century, Charles Burney – a musician, travelogue writer, and historian of music – travelled across Europe in the early 1770s. While there, he recorded observations, not just on the state of music, but on society as a whole. Travelling through the Holy Roman Empire in 1772, he observed at Württemberg: 'The soldiers seem disciplined into clock-work. I never saw such mechanical exactness in animated beings. One would suppose that the author of *Man a Machine* had taken

---

1 LOC: Washington Papers, George Washington to John Hancock, 25 September 1776.
2 Anon., *Proposals for Raising a Standing Army of Wax* (Dubin: Theo Jones, 1733), pp.2–5.

his idea from these men: their appearance, however, is very formidable.'[3] Throughout the eighteenth century, observers and officers noted the tighter and tighter levels of control across the armies of military Europe.

These stories capture two opposing schools of thought in eighteenth-century military Europe. The first poked fun at the idea of transforming men into machines as patently ridiculous. The second believed that the results of mechanical discipline were impressive. For eighteenth-century military men, this paradigm proved difficult to escape from during the entire period covered by this book. In 1730, the idea of a machine army propelled by clockwork was humorous. By 1783, turning men into unfeeling biological machines had become a real desire on the part of a significant number of officers. As one Saxon officer, Johann Gottfried von Hoyer, put it: 'This is embarrassing, as after one hundred years of practice, we cannot bring common soldiers under control, and build a robotic shooting-machine.'[4] The language of the soldier as a 'shooting machine' (*Schiessmaschine*) was common in German-language texts throughout the era, often with a negative or unfeeling connotation.[5]

In the northern Italian battlefields of 1734, we saw two armies who embraced a style of fighting which made sense to the infantry soldiers who took part. These infantrymen fired at will, took cover from enemy fire, utilized terrain effectively, and moved at speed on the battlefield. This is clearly at odds with popular memory of eighteenth-century warfare in the present. It is clear, then, that this more intuitive style of fighting must have been challenged by another form of infantry warfare, one which emphasized rigid control over individual initiative. In the period between 1733 and the present, a discourse emerged which prioritized the tactics of control. This version of eighteenth-century battle emerged in response to the success and perceived tactics of Frederick II of Prussia, during the War of Austrian Succession and Seven Years War.

In the War of Austrian Succession (1740–1748), Frederick II's army won a string of impressive victories over its Austrian and Saxon opponents. Overwhelmingly, this was due to the impressive performance of the Prussian infantry on the battlefield. At Mollwitz in April 1741, Prussian infantrymen won a battle that appeared so lost that Frederick II had already abandoned the battlefield. At Chotusitz, Hohenfriedberg, Soor, Kesselsdorf, and a host of smaller actions, the Prussian army defeated more numerous enemy armies. European military observers, followed to some degree by King Frederick himself, came to believe that this spectacular run of successes was a result of tightly controlled infantry who fought in a uniform and disciplined manner. As a result, by the early 1750s across the continent, European military writers became obsessed with the idea of tightly controlling their soldiers, in order to reproduce the supposed Prussian formula of victory. A generation of Prussomaniacs was born.

---

3   Charles Burney, *The Present State of Music in Germany, The Netherlands, and United Provinces* (London: T. Beckett and Co., 1773), vol.1, p.103.
4   Johann Gottfried von Hoyer, *Geschichte der Kriegskunst* (Göttingen: Rosenbusch, 1799), pp.102–103.
5   Andreas Georg von Rebmann, *Wanderungen und Kreuzzüge durch einen Theil Deutschlands* (Altona: Verlaggesellschaft, 1795), p.49. Gustav Erichson, *Das Neue Graue Ungeheuer* (Uppsala: Unknown, 1795), p.30.

Frederick II's army was knocked about quite badly in the Seven Years War (1756–1763), and Frederick himself was forced to adapt many of his assumptions regarding the nature of tactics and infantry battle. Despite a harrowing experience for Prussia, the Prussian army again performed well, to some degree because of Frederick's flexibility. This led to legendary status for the Prussian king and his army. For the remainder of Frederick's life, military men from across Europe and the world flocked to Berlin, Potsdam, and Breslau to see the Prussian army during the semi-realistic wargames and manoeuvres the Prussians carried out each year. The robotic rigidity and iron discipline of the eighteenth-century Prussian army were vastly overblown in the minds of European contemporary officers. Overwhelmingly, the popular twentieth- and twenty-first-century image is drawn from what Prussomaniac officers wanted – that is to say, from their hypothetical prescriptive thoughts about eighteenth-century warfare. The period between 1733 and 1783 saw an ever-increasing amount of literature from Prussian-influenced officers, detailing what tactics they believed would lead to the most success on the battlefield.

This caused a series of key tactical debates in eighteenth-century military circles. Should officers allow their men more freedom to engage the enemy as they wish? Should skirmishers be allowed to range out ahead of the battalion on their own initiative? Should the men be allowed to walk at a natural pace, or was regulation of the step required? Should soldiers be allowed to fire their weapons freely, or should they be under strict orders not to fire? A writer and politician associated with the Northamptonshire militia posed the argument this way:

> Let us not despise our militia as unfit for our protection. Though they be defective in military discipline… I should think them able to cope with the best regular troops… personal courage, an implacable hatred of the enemy, and a mutual confidence between the officers and men of a newly-raised army have often defeated the mechanical motions of well-disciplined men.[6]

In their own way, these questions formed literature as important as the classic debate initiated by the chevalier de Folard over the use of column versus linear tactics. In modern popular memory, as well as in the literary war in their own time, the Prussomaniacs were victorious, obscuring an important strain of eighteenth-century military thought and practice that disagreed with them.

The Prussians in Myth and Reality

Eleazar Mauvillon did nothing but echo the thoughts of a multitude of European observers when he commented on Prussian troops in 1756: 'The fervour with which they have been exercised without ceasing for the last thirty years has accustomed them to make their evolutions and movements with admirable concert. A body of

---

6   John Hope, *Hope's Curious and Comic Miscellaneous Works* (London: Unknown, 1780), pp.175–176.

Prussian infantry is like a machine who officers direct its actions.'[7] English, French, Spanish, Sardinian, and even American soldiers travelled to Prussia in order to observe the Prussian military machine in its natural environment. Even before the age of Frederick II of Prussia, English travellers associated the kingdom and its military with discipline. Alexander Dury, a young officer of the guards who accompanied King George II on a tour through the courts of Europe in 1732, made particular note in his diary on arriving to Potsdam:

> Went to the parade at Potsdam, where the king was present and the Prince Royal, the parade consisted of 300 men, two battalions of the great grenadiers in Potsdam of 700 men each, besides those there are near 700 supernumerary Grenad[iers]. They are very regular in all their motions, very clean about their arms and dress, and very strict in their discipline.[8]

British descriptions of Prussian discipline took on a life of their own in the later eighteenth century. If some Britons, such as Dury above, noted it in passing, others devoted whole paragraphs to the inner workings of obedience in Prussia. For many British military men and civilian intellectuals, the subject became one of almost morose fixation. The travelogue author John Moore asserted that, 'in Germany, where the passions are annihilated... a man is modelled into a machine before he is thought a good soldier.'[9] John Moore devoted no fewer than three chapters of his travelogue to describing 'Prussian Discipline,' going so far as to include conversation with a Prussian officer on the subject. In this possibly fictional dialogue, Moore observed a soldier being beaten and pressed the officer to explain why the Prussian military system was so cruel. The officer responded, 'Everything must be considered as of importance by a soldier... which his officer orders him to do.' The unnamed Prussian continued: 'In all probability, the fault was involuntary, but it is not always possible to distinguish involuntary faults from those that happen through negligence. To prevent any man from hoping that his negligence will be forgiven as involuntary, all blunders are punished.'[10] Moore continued on this theme for pages, using three different types of examples of unjust punishment to attempt to persuade his Prussian acquaintance of exaggerated discipline, but the Prussian did not relent.

Many Anglophone observers connected Prussian military discipline to a supposed lack of morale among the Prussian Army and a propensity among Prussian soldiers to desert. American Colonel Smith noted that Frederick's 'military abilities are undoubtedly great and had he the affections of his army he might be a second conqueror of the world – his armies are composed of dissatisfied mercenaries, compelled by severity of discipline to discharge their duty.'[11] Johnson's trans-

---

7 Eleazar Mauvillon, *Histoire de la Dernière Guerre de Boheme* (Amsterdam: Mortier, 1756), p.102.
8 NAM: 1986-12-38-116, diary of Alexander Dury.
9 John Moore, *A View of the Society and Manners in Italy* (Dublin: Price, 1781), p.152.
10 Moore, *A View of Society and Manners in France, Switzerland, and Germany*, vol.2, p.160.
11 Fransico de Miranda, *Archivo del General Miranda* (Caracas: Parra Leon Hermonas, 1929–1950), vol.1 p.383.

lated work notes that though the military strength of Prussia might appear great, 'this is perhaps not so much owing to the intrinsic bravery of his troops, as to the excellence of their discipline.'[12] John Burgoyne noted, 'The king of Prussia deprived of such principles to work upon, turns his defects to advantage, and substitutes a species of discipline wherein the mind has no concern.'[13] Robert Jackson believed that 'the Prussian soldiery may be supposed, in consequence of a rigid drilling, to have acquired certain mechanical habits, at the commencement of a war, which were capable of giving them advantages in action over a less experienced enemy.'[14] James Boswell visited Prussia just after the Seven Years War and observed a Prussian regiment drilling. Boswell recorded that 'the soldiers seemed in terror. For the least fault they were beat like Dogs.'[15] Far from believing this hindered the Prussian army, however, Boswell continued, 'I am, however, doubtfull if such fellows don't make the best soldiers. Machines are surer instruments than men.'[16] This presentation of soldiers as machines formed an enduring stereotype.

Observing the Prussian army in 1766, John Burgoyne (famous in American history as the man defeated at the Battle of Saratoga) railed against the Prussian service. Burgoyne claimed, 'the ranks are filled up, perhaps more than a third part, with strangers, deserters, prisoners and enemies, of various countries, languages, and religions.'[17] His account continued, 'They cannot therefore be actuated by any of the great moving principles which usually cause extraordinary superiority in armies; they have neither national spirit nor attachment to their prince, nor enthusiasm'.[18] Burgoyne concluded, 'The first principle of the Prussian system is subordination, and the first maxim 'not to reason, but to obey.'[19] Burgoyne's key point, though, is important to consider. He believed that because many Prussian soldiers had no connection to place, that they could not be motivated by 'national spirit, attachment to their prince, nor enthusiasm.'[20]

Another British officer who served in America during the War of Independence, Robert Jackson, came to different conclusions regarding the motivation of Prussian common soldiers. Mere discipline was not enough to explain the experiences of eighteenth-century soldiers, as Ilya Berkovich, Michael Sikora, and Sascha Möbius have all recently argued.[21] Jackson, a medical officer in the British 71st Regiment of

---

12 Jacques-Antoine-Hippolyte, comte de Guibert and Joseph Johnson (trans.), *Observations on the military establishment and discipline of the King of Prussia* (London: Fielding and Walker, 1780), p.3.
13 Edward Fonblanque, *Political and Military Episodes in the Latter Half of the Eighteenth Century* (London: MacMillan, 1876), p.65.
14 Robert Jackson, *A Systematic View of the Formation, Discipline, and Economy of Armies* (London: Stockdale, 1804), p.68.
15 James Boswell, *Journal of his Swiss and German Travels* (Edinburgh: Edinburgh University Press, 2008), reprint, p.95.
16 Boswell, *Journal of his Swiss and German Travels*, p.95.
17 Fonblanque, *Political and Military Episodes in the Latter Half of the Eighteenth Century*, p.65.
18 Fonblanque, *Political and Military Episodes in the Latter Half of the Eighteenth Century*, p.65.
19 Fonblanque, *Political and Military Episodes in the Latter Half of the Eighteenth Century*, p.65.
20 Fonblanque, *Political and Military Episodes in the Latter Half of the Eighteenth Century*, p.65.
21 Ilya Berkovich, *Motivation in War: The Experience of Common Soldiers in Old-Regime Europe* (Cambridge: Cambridge University Press, 2017), Michael Sikoura, *Discinplin und Desertion:*

Foot, reached this conclusion in the late eighteenth century, and his observations remain fascinating. Jackson wrote, 'Something is still wanting to account for the conduct of the Prussian troops, which rose above anything that can be expected to arise from an impulse of fear or coercion.'[22] In Jackson's view,

> Heroic acts do not originate in fear: yet the acts of the Prussian soldiers were often heroic... The native Prussian, irritated and inflamed against the invaders of his country (for the peasant venerates the soil which protects the ashes of his fathers), became ardent in defence; the old soldier was proud of his renown; the recruit, whether forced or trepanned, was carried away in the torrent; he imbibed a sensation of glory from anticipated success, – and became proud in himself.[23]

Jackson's poetic musings are romantic in the extreme but stand as almost completely antithetical to Burgoyne's argument. Jackson believed that because Prussian soldiers were connected to place, they were motivated by something higher than fear and coercion. John Moore also noted the importance of family connections in the lives of common soldiers in the Prussian army.[24] David Dundas, a British officer trying to reform the army after the American War of Independence, noted that 'the native part of the Prussian Army is composed of a better kind of man, than that of any other nation.'[25] Thus, Dundas refers to a 'better kind of man' to indicate that as a result of conscription Prussian soldiers came from a different section of society than other armies. Dundas also noted that foreigners in the Prussian service, 'many of them from marriage, habit, etc, are attached to the country.'[26] Even the foreign soldiers, Dundas indicates, are in the army as a result of more than mercenary reasons. Although British commentators failed to reach a firm consensus regarding the Prussian army and its motivation, most writers agreed with the stereotype that discipline outweighed the nativist motivations available to part of the army. These British authors also observed, however, that Prussians were ruled by a despotic monarch, and guarded by rigidly disciplined soldiers.

In addition to the seemingly despotic nature of the Prussian military state, many military writers cried out against the Prussomaniac tendency to copy Prussian uniforms and drill. An officer from Wales who served in many European armies, Henry Lloyd, railed:

> Short cloaths, little hats, tight breeches, high-heeled shoes, and an infinite number of useless notions in the exercise and evolutions have been

---

Strukturprobleme militärischer Organisation im 18. Jahrhundert Historische Forschungen 57 (Berlin: Duncker und Humboldt, 1996), Sascha Möbius, *Mehr Angst vor dem Offizier als vor dem Fiend? Eine mentalitätsgeschichtliche Studie zur preußischen Taktik im Siebenjährigen Krieg* (Saarbrücken: VDM Verlag, 2007).
22 Jackson, *A Systematic View of the Formation, Discipline, and Economy of Armies*, p.70.
23 Jackson, *A Systematic View of the Formation, Discipline, and Economy of Armies*, p.70.
24 Moore, *A View of Society and Manners in France, Switzerland, and Germany*, vol.2, p.209.
25 BL: King's Manuscripts 241, p.36.
26 BL: King's Manuscripts 241, p.36.

introduced, without any other reason than their being Prussian; as if really these things could contribute to gain one battle, make a fine march or manoeuvre, carry on the operations of siege, choose a fine camp or position, etc.[27]

For Lloyd, this was Prussomania at its height: a focus on the uniforms that 'they are so infatuated with.' He continued, 'I have no great regard for men who are attached to such trifles,' and he would be 'very indifferent to the opinion they may be pleased to form of me.'[28] He was far from the only author to find the focus on the details of uniforms, dress, and outward appearance of soldiers a little annoying.

On his 1773 trip to Prussia, the comte de Guibert was observant enough to note of the Berlin garrison:

> The soldiers are clean and well kept, but without any affectations; for example hair powder. How ridiculous are those in France who think they are imitating Prussian uniforms. In Prussia there are no frills, no pompoms, no affectations, no uniformity. Some regiments have red collars, some have black collars, some have hats, others copper caps; the cut of the clothes varies between regiments: they are as they were first instituted. We were blissfully unaware of the wise indifference of the King of Prussia to all these mere externals.[29]

He continued, 'the infantry makes a pleasing appearance, looks well under arms, for the ease of the soldiers, and above all the ease of the march. No artificial rigidity.'[30] Watching a battalion manoeuvre, he commented, 'they returned to the march, pace free, quite long, quite fast, quite natural.'[31]

Foreign officers who observed the Prussians were sometimes surprised to see their expectations failed to live up to reality. Upon observing a Prussian battalion manoeuvring in 1774, David Dundas recalled that the men 'did not keep the step, but were lengthened out,' and that the Prussians 'fire kneeling.' In 1785, he recalled, 'there are many absurd practices in this service as well as in every other – the king has his whims which he preserves in, among these may be reckoned… the front rank kneels.'[32]

One expectation that did seem to live up to the myth is the repeated comment, 'load[s] quickly,' 'load[s] very quick,' 'prodigious speed in loading.' The comte de Guibert repeats the phrase, 'prodigious speed in loading,' three times in his journal.[33] Officers observed that Prussian troops were 'steady under arms,' and that

---

27 Henry Lloyd, *Continuation of the History of the Late War in Germany* (London: Hooper, 1791), p.ix.
28 Lloyd, *Continuation of the History of the Late War in Germany*, p.ix
29 Guibert, *Journal d'un Voyage en Allemagne en 1773* (Paris: 1803) p.165.
30 Guibert, *Journal d'un Voyage en Allemagne en 1773* (Paris: 1803) p.179.
31 Guibert, *Journal d'un Voyage en Allemagne en 1773* (Paris: 1803) Volume 2, p.128.
32 BL: King's Manuscripts 241, p.12.
33 BL: King's Manuscripts 240, p.5.; Guibert, *Journal d'un Voyage en Allemagne en 1773*, vol.1, pp. 207, 224; vol.2 p.128.

'their attention and exactness is very great.'[34] David Dundas noted upon reviewing the Prussians in the 1780s:

> The great exactness of the Prussian march and of their correct lining on all occasions depends on the three circumstances. The sameness of the step to which each individual is habituated – the never quitting the touch of the man to whom they are to dress – and the perfect squareness of the shoulders and the body. These combined ensure the justness of the line, by a glace of the eye, and without turning the head.[35]

Dundas was convinced of the idea that a slow and orderly march was more effective than a quick march which led to disorder. He felt compelled to defend this idea at length, arguing:

> Altho the general movements of the Prussian infantry appear slow and solemn, yet they are so accurate that no unnecessary time being lost in dressing or correcting distances, they arrive sooner at their object than any others, and at the instant of forming they are in perfect order to make the attack. The quickness and decision of officers of all ranks is very striking, and the diligence, and the dispatch of all staff officers who are to give motion to the machine is very conspicuous.[36]

Here, Dundas employs the language of the battalion as machine, not individual soldiers – but the unit as a whole. For him, initiative and decision-making were vital, but on the part of the officers, rather than the men, who were to be directed as component parts of a larger mechanical whole. American officers were not immune to the charms of Prussomania. Colonel William Stevens Smith, the son-in-law of John Adams, visited Prussia in the 1780s and was astounded by what he saw:

> The troops are superior to panegeric – they marched in platoons from the town to their ground, on the word, the line was formed with the greatest perfection – they advanced in line, advanced by platoon firing, and then by regiment, and finally, on a supposition of being pressed by cavalry, they gave ground, retiring in eight detached lines in such a manner as to perfectly cover each other.[37]

Smith had many negative comments on the nature of Prussian society but could not help but be impressed with their infantry and noted, 'the troops manoeuvre with great ease – the display of the column and the advance in line is astonishing.'[38] Although, in Smith's mind, a horrifying society had produced the Prussian military

---

34 BL: King's Manuscripts 240, p.5.
35 BL: King's Manuscripts 241, p.12.
36 BL: King's Manuscripts 241, pp.14–15.
37 Miranda, *Archivo del General Miranda*, vol.1, p.380.
38 Miranda, *Archivo del General Miranda*, vol.1, p.398

system, its machine-like exactness still left room for admiration on the part of the American officer. Oddly, Smith does not make any reference to Friedrich Wilhelm, Freiherr de Steuben, who had Prussianized his own army to some degree.

Across military Europe, officers were directly inspired by Prussia as they attempted to devise better systems by which to tactically direct their armies. The vast majority of officers believed that the best way to do so was to bring soldiers more directly under control and eliminate any possibility of soldiers exercising their own control over the management of the battle. As officers and soldiers marched toward, fired on, and charged the enemy throughout the eighteenth century, many officers dreamed of ways that they could bring their soldiers more directly under control, to build the 'robotic shooting machine' envisioned by Johann Gottfried von Hoyer.

What Officers Wanted: Marching

This book has already explored the incredible impact of cadenced marching on pre-battle deployment: allowing large number of infantry to simultaneously deploy in a quick and efficient manner. This was especially true when pre-battle deployment called for complicated manoeuvres, such as the Prussian *deployiren*. Before formations were disrupted by enemy artillery and small arms fire, and before the men fell into disarray as a result of their desire to fire back at the enemy, cadenced marching was an effective tool which allowed its practitioners, like the Prussians, to maintain a decisive advantage in the manoeuvring just before the fighting began. Other officers wanted to take this further: they fanaticized about creating infantry battalions that would never fall into disorder or need to be re-aligned. The widespread re-emergence of cadenced marching seemed to promise new possibilities in this regard. Once again, the officers of military Europe looked to Prussia, hoping that the successes that the tightly controlled Prussian military could be reproduced in other armies and locales. By marching at a particular pace, perhaps guided by the music or the beating of drums, officers hoped the men would always be in order, saving valuable time as the units marched across the battlefield in combat.

During his 1773 tour in Prussia, the French comte de Guibert noted, 'The leading principle of his Prussian majesty's military system, seems to be to reduce his troops to the nature of machines, to teach them to have no will of their own, and to be as deaf and pitiless as their muskets.'[39] Even officers before the rise of European Prussomania believed that tightly controlled soldiers should be heavily regulated on the battlefield. British Lieutenant General Humphrey Bland exhorted in his treatise: 'In marching up to attack the enemy, the line should move very slow, that the battalions may be in order, and the men not out of breath when they come to engage.'[40] Keeping the battalion in order was the paramount desire of many eighteenth-century infantry officers; it would save them from sudden surprises like the onset of cavalry attacks.

---

39 Guibert and Johnson, *Observations on the military establishment and discipline of the King of Prussia*, p. 85.
40 Humphrey Bland, *A Treatise of Military Discipline* (London: Daniel Midwinter, 1743), p.134.

Another French officer, the marquis de Toulongeon, observing the Prussian manoeuvres at Potsdam in the 1780s, described the order of the Prussian infantry as 'this perfection,' and believed that it stemmed from 'starting off on the left foot, resuming the march of 75 paces per minute. The measure of this step is so imprinted on the soldiers' mind and his legs are so used to the pace that it is like watching the working of a pendulum.'[41]

British commander James Wolfe noted to Lord George Sackville in 1758:

> When soldiers are the masters of the use of their firearms and of their bayonets, the next great object is their marching in battalion, as your Lordship knows full well. For this, no good instructions have ever been given in my time, nor any principles laid down by which we might be guided. Hence the variety of steps in our infantry, and the feebleness and disorderly floating of our lines. General Drury, I think, has the merit of the late inventions; 'tis unlucky, however, that our great master in the art of war, Frederick of Prussia, was not preferred upon this occasion. He has made the exercise simple and useful; we cannot choose so good a model.[42]

In Wolfe's mind, variety was the source of the problem, and a singular step would have prevented the 'disorderly floating' that officers so feared. Although particularly associated with the seemingly robotic Prussians, the slow-cadenced march was being experimented with across Europe. Vincent-Juddes, marquis de Saint-Pern, demonstrated cadenced marching for the French army in 1758:

> M. de Saint-Pern, Lt. General, who led this column, no doubt wishing to show-off to the whole army, made this column march to the rear in order. The column was in fine order, keeping exactly its ranks and files. What did it matter that the slowness delayed its arrival? Its drums beat on the field, and the column followed the pace of the cadence slowly and rhythmically.[43]

The comte de Guibert was convinced of the importance of regulating the cadence and wrote at length on the subject. He sought to answer the question, 'why not abandon the soldier to his free step, to the step that he uses outside of school [of the solder]?' Guibert replied on the importance of a 'measured [and] cadenced step whose form and speed are absolutely common to all the legs of a battalion.' After dwelling on the natural differences in the steps of men from different regions, Guibert returned to the importance of 'mechanical bending,' 'the mechanism of the march,' and concluded, 'I was obliged to subject the soldier for the manoeuvre march to a uniform step because this march must be

---

41 Marquis de Toulongeon, *Une Mission Militaire en Prusse en 1786* (Paris: Imprimeurs de l'Institut, 1881), pp.193–194.
42 Wilson, *The Life and Letters of James Wolfe*, p.255.
43 Jacques de Mercoyrol de Beaulieu, *Campagnes de Jacques de Mercoyrol de Beaulieu, capitaine au régiment de Picardie: 1743–1763* (Paris: Libairie Renouard, 1915), p.184.

together and precise… one must be particularly attached to establishing in the soldiers the equality of step.'⁴⁴

British officer Thomas Simes, writing in the 1780s, exhorted his reader regarding the importance of maintaining order at length. Failure to govern the men's steps carefully, in Simes's view, led to a battalion that 'improperly conveys the idea of a machine, constructed upon no principle, which is ready to fall in pieces every moment'. Continuing his hypothetical description of the march, Simes noted, 'if, on a march, the front is ordered to quicken its pace, the rear must lose ground… which present throws the whole corps into disorder.' Disorder, the great enemy, had reared its ugly head. Simes continued, '…it becomes impossible to march a body of troops with expedition, without forsaking all manner of order and regularity.' Fortunately, Simes believed that he had the answer to this seeming conundrum.

> It is nothing more than to march in cadence… by means of this, you will always be able to regulate your pace at pleasure; your rear can never lag behind, and the whole will step with the same foot; your wheeling will be performed with celerity and grace; your men's legs will never mix together, you will not be obliged to halt… in the middle of every wheel to recover the step.⁴⁵

For Simes, cadenced marching, specifically to the sound of music, was the method to preserve all discipline. One imagines he might have been concerned about reports that Prussian troops did not use music to manage their cadence. David Dundas recalled upon observing Prussian troops: 'Except in parade, music is very sparingly used in marching, and never with a view of assisting the regularity of the step; but from time to time the drums give a roll and particularly when the word march is given.'⁴⁶ Only by imprinting a march pace 'on the mind' of the soldier, as the marquis de Toulongeon argued, could officers hope to match the disciplined Prussians.

In observing the placid drill-square marches of these Prussians and their imitators, we are a long way from practice of the French in the battles of 1734, or even the pace used by the British in the American War of Independence, to say nothing of the Swedes during the Great Northern War. There is also a great gulf between what the Prussomaniacs observed in the 1770s and 1780s, and what was actually performed on the battlefield of the Seven Years War, as the following chapters will show. Having described the slow orderly marches carefully observed and desired by Dundas, Toulongeon, and Bland, the chapter will turn to examining what these officers desired in the realm of firepower.

What Officers Wanted: Firepower

Across military Europe, many officers were horrified by the natural tendency of soldiers to fire independently of their officers' orders. This was a problem (in their

---

44   Guibert and Abel, *Guibert's General Essay on Tactics*, pp.62–63.
45   Thomas Simes, *A Treaty of Military Discipline* (London: John Millan, 1780), p.249.
46   BL: King's Manuscripts 241, 11.

minds) that ran across all major militaries of the eighteenth century. The tendency of men to fire without orders created several major crises of confidence for their officers. First, the natural inaccuracy of the smoothbore flintlock weapon meant that these weapons were most effective when under 100 yards from the enemy. As a result, some officers hoped that the men would preserve their fire into a range where it was most effective. The first volley of musketry was critical, as it was delivered with a clean, ready weapon unlikely to misfire. As a result, some officers, particularly in the British Army, believed that the most effective way to deliver fire was by closing to a close range before beginning to fire. Second, once a battalion opened fire, it became difficult for the officers to get that battalion to do anything else. Any forward advance might be stalled, as the men kept firing without orders as soon as they had loaded. Third, officers worried that if the entirety of the battalion fired at once, at too great a distance from the enemy, the battalion would be devoid of fire, and could be quickly overrun by enemy cavalry forces.

In 1734, Austrian *Major* Johann Maquire recalled Prince Joseph von Saxe-Hildburghausen's orders: 'the Prince ordered me to gallop from company to company and forbid on pain of death to fire without orders from him or me.'[47] Advancing to close range before giving fire was an aspirational ideal that many officers strove to achieve. In one of the most frequently reprinted British military treatises of the century, Lieutenant General Humphrey Bland warned his readers: 'The commanding officer of every battalion should march up close to the enemy, before he suffers his men to give their fire.'[48] Bland continued: 'it being a received maxim, that those who preserve their fire the longest will be sure to conquer.' Again and again, Bland argued, 'a battalion of foot would manage their fire to the best advantage, [to] not throw it away at too great a distance, which they are apt to do.'[49] Bland concluded:

> In advancing towards the enemy, it is with great difficulty that the officers can prevent the men (but more particularly when they are fired at) from taking their arms without orders, from their shoulders, and firing at too great a distance. How much more difficult must it be to prevent their firing, when they have their arms in their hands ready cock'd and their fingers on the trickers? …Therefore in my opinion [it is] imprudent to trust a thing of this consequence to chance, unless obliged to it by having no other means left.[50]

Bland's comments on the importance of leaving muskets shouldered were echoed by many military authorities, such as Frederick II and George Washington. Indeed, many authorities seem to have preferred holding fire until at very close range. David Blackmore has argued that this was specifically a British doctrine, but it was shared

---

47 Historical Manuscripts Commission, *Report on Manuscripts in Various Collections*, vol.8, p.408.
48 Bland, *A Treatise of Military Discipline*, p.134.
49 Bland, *A Treatise of Military Discipline*, p.91.
50 Bland, *A Treatise of Military Discipline*, p.80.

by officers in the American, Swedish, French, and Prussian Armies.[51] A French officer writing early in the century asserted, 'We must train the soldiers above all to hold their fire, and to endure the fire of the enemy. In normal circumstances a battalion is beaten when once it has opened fire, and the enemy still has its fire in reserve.'[52] Some of the most well-known advocates of this doctrine were British officers, and the Americans influenced by them.

James Wolfe, as a young officer with the 12th (Durore's) Regiment of Foot at the Battle of Dettingen in 1743, recalled the fight between two bodies of infantry there. In a letter to his father after the battle, he remembered, 'The Major and I (for we had neither Colonel nor Lieutenant-Colonel) before they [the French Infantry] came near, were employed in begging and ordering the men not to fire at too great a distance, but to keep it till the enemy should come near to us.'[53] Wolfe's language of begging matches the previous descriptions of threats, cajoling, and pleading that other officers such as the Austrian Johann Maquire employed when trying to prevent the men from opening fire too early.

Throughout his entire military career, George Washington maintained a firm and relatively traditional approach to tactical firepower doctrine. As an officer and a man, Washington came of age during the French and Indian War, when, in theory at least, British tactical doctrine called for holding fire until at extremely close range, perhaps under 30 yards.[54] Washington, in the course of his military writings and career, never wavered from the belief that holding fire to within very close range, and above all, never firing without orders, was the correct way to employ infantry fire.

On 20 November 1758, as a young officer in the British provincial army during the Seven Years War, Washington outlined an early version of the standard British tactical doctrine: 'In case of action it is particularly recommended to each officer to keep their men calm and prevent them throwing away their fire which they are too apt to do when in a hurry.'[55] The elements of the tactical doctrine he embraced – for the men to remain calm, and never throw away their fire – would remain with him for the rest of his military career.

Upon taking control of the young Continental Army in 1775, Washington immediately set about mandating his tactical thoughts into doctrine. On 9 January 1776, Washington reminded his troops in their general orders, 'nothing betrays greater signs of fear, and less of the soldier, than to begin a loose, undirected and unmeaning fire, from which no good can result, and nor any valuable purpose answered.'[56] On 3 March 1776, Washington's general orders reminded officers to be 'so posted as to keep the men to their duty; particular care is to be taken to prevent their firing at too great a distance, as one fire well aimed does more execution than a dozen at

---

51 David Blackmore, *Destructive and Formidable: British Infantry Firepower, 1642–1765* (London: Frontline Books, 2014), pp.167–173.
52 Charles Sevin de Quincy, *Histoire militaire du règne de Louis le Grand* (Paris: Imprimeur du Roi, 1726), vol.8, p.67.
53 Wilson, *The Life and Letters of James Wolfe*, p.37.
54 Blackmore, *Destructive and Formidable*, pp.167–168.
55 LOC: Washington Papers, Orderly Book Entry, 20 November 1758.
56 LOC: Washington Papers, General Orders, 6 January 1776.

long shot."⁵⁷ Washington would return to his insistence that men avoid firing at long range again and again in his general orders, giving a clear indication that he was not pleased with the army's performance in this regard.

On 29 June 1776, Washington gave his clearest articulation of his tactical doctrine:

> All soldiers… will behave with coolness and bravery, and will be careful not to throw away their fire. The General recommends them to load for their first fire with one musket ball and four or eight buckshot according to the size and strength of their pieces. If the enemy are received with such a fire at not more than twenty- or thirty-yards distance, he [Washington] has no doubt of their being repulsed.[58]

When it came to the utilization of infantry firepower, this would remain Washington's clear formula for victory: 1) soldiers should remain calm, 2) soldiers should never fire without orders, 3) soldiers should hold their fire until the enemy is within very close range.

Though often remembered for his antagonistic relationship with Washington later in the war, in 1776, Charles Lee was in lockstep with his commander on matters of tactical doctrine. Writing to Colonel William Thompson, commander of a battalion of riflemen, Lee railed on this issue:

> It is a certain truth that the enemy entertain a most fortunate apprehension of American riflemen. It is equally certain that nothing can diminish this apprehension so infallibly as a frequent ineffectual fire. It is with some concern, therefore, that I have been informed that your men have been suffered to fire at a most preposterous distance, Upon this principle, I must entreat and insist, that you consider it as a standing order, that not a man under your command is to fire at a greater distance than one hundred and fifty yards, at the utmost; in short, they must never fire without almost a moral certainty of hitting their object. Distant firing has a doubly bad effect; it encourages the enemy, and adds to the pernicious persuasion of the American soldiers, viz: that they are no match for their antagonists at close fighting. To speak plainly, it is almost a sure method of making them cowards. Once more, I must request that a stop be put to this childish, vicious, and scandalous practice.[59]

While Lee's exact prescription for range changed as a result of the riflemen he addressed, his complaint was much the same as Washington's to the rest of the army. At the Battle of Camden in 1780, Horatio Gates likewise instructed his men to fire at extremely short range – in this case, six paces.[60]

---

57   LOC: Washington Papers, General Orders, 3 March 1776.
58   LOC: Washington Papers, General Orders, 29 June 1776.
59   Peter Force, *American Archives*, Fifth Series: 1776 (Washington: St. Claire and Force, 1848), vol.1, p.99.
60   NARA: M804, , Pension application of Thomas Moody. W25732.

As the war stretched in its middle years, Washington's pronouncements regarding short-range fire became less frequent but more emphatic. In a series of contingency orders for a potential battle near West Point in 1779, Washington reminded his troops: 'It is expected that the troops will advance boldly upon the enemy – and by no means, nor under any pretense whatsoever throw away their ammunition at long shot – A musket had better never be discharged than fired in so wasteful, shameful, & cowardly a manner.'[61]

British officers also reproved their troops for similar offenses in their general orders. After the Battle of Freeman's Farm in September of 1777, Lieutenant General John Burgoyne noted:

> Amidst these general subjects of applause, the impetuosity and uncertain aim of the British troops in giving their fire… is to be much lamented. The Lieut. General is persuaded that this error will be corrected in the next engagement, upon conviction of their own experience and reason, as well as upon the general principle of discipline *never* to *fire* but by the *order of an officer.*[62]

During the Seven Years War, British officers had similar fears. William Windam and George Townshend argued in a treatise for instructing militia troops that while firepower was important, the main concern should be that 'the battalion is not so much broken, and is easier to keep in order.'[63] Firepower itself had taken a subordinate position to keeping the men tightly under control and in order.

In the period between 1733 and 1783, many officers hoped that by controlling their soldiers and marching them to within close range of the enemy before firing, they would gain a decisive advantage, driving the enemy with a deadly opening volley. Although this was a particularly cherished belief in the British Army, other armies continued to believe that this doctrine should be emulated. Holding fire until within close range, these officers supposed would enable them to drive off the enemy more quickly during the truly decisive phase of the battle: the charge.

What Officers Wanted: Charging

During the eighteenth century, the officers of military Europe wrote in a distinctly neo-classical paradigm. Inspired by their cultural familiarity with classical antiquity, officers sought to emulate the successes of the Greeks and Romans. Many military treatises contain classical references; others contain whole sections devoted to explaining classical tactics, and why they were useful

---

61 LOC: Washington Papers, Contingency Orders: General Disposition for the Army, 12 June 1779.
62 Edward O'Callaghan, *Orderly Book of Lieut. Gen. John Burgoyne* (Albany: Munsell, 1860), p.116. Emphasis in original.
63 William Windam and George Townshend, *A Plan of Discipline: Composed for the Use of the Militia* (London: Shuckburgh, 1760), p.18.

and relevant in an eighteenth-century military context. An admiration for explaining the 'military art' of the Romans and Greeks led these men to dwell heavily on the importance of charging into close contact with enemy forces and defeating the enemy decisively in melee fighting. There were many calls to readopt the pike as a force in military Europe, stemming from figures as diverse as the chevalier de Folard and Benjamin Franklin. For officers who believed that firearms represented a weaker form of the military art, the answer remained to break through the enemy position with boldness, as Charles XII had so often done in the earlier part of the century.

Henry Lloyd, a military adventurer who served in a number of different European armies, declared in an essay posthumously published in 1784:

> Let us… compare the advantages and defects of missile and hand weapons… Fire-arms are calculated for a defensive war and to keep the enemy at a distance… but are of no use when he can approach you. Hand-weapons… can be of no use at a distance, but are necessary when the armies approach each other. The effects of the one are precarious and undecisive; those of the latter certain and complete. The musket is the resource of prudence and weakness; hand-weapons are the arms of valor and vigor.[64]

Lloyd's belief in the indecisive nature of firefights led him to the only, in his mind, reasonable conclusion: 'It follows, therefore, that a certain number of men should be armed with pikes.' Not content to turn back the clock to the 1690s, he also advised re-issuing infantry armour. Lloyd believed that armour was 'always useful, it gives confidence to the men and likewise diminishes and sometimes destroyed entirely the effect of a musket-ball when fired from a certain distance.'[65] Throughout the century, military authors continued to believe that decisive shock action would return to the battlefield, replacing the firefight as the primary method of dislodging and defeating the enemy.

Frederick the Great's views on infantry combat changed throughout his military career, as Christopher Duffy has charted.[66] After the war of Austrian Succession, Frederick argued it would be best for infantry to attack without firing. Describing a hypothetical attack on an enemy position, he argued:

> On these occasions, I would not permit the infantry to fire, for it only retards their march, and the victory is not decided by the number of slain but the extent of the territory which you have gained. The most certain way of ensuring victory is to march quickly and in good order against the enemy, always endeavoring to gain ground.[67]

---

64 Lloyd, *Continuation of the History of the Late War in Germany*, p.17.
65 Lloyd, *Continuation of the History of the Late War in Germany*, pp.36–37.
66 Duffy, *The Army of Frederick the Great* (2020), p.148.
67 Frederick II and T. Forster, *Military Instruction from the Late King of Prussia to His Generals* (Sherborn: J. Cruttwell, 1818), p.114.

He continued to outline similar thoughts in his *Principes Géneraux de la Guerre* of 1748, saying, 'the infantry will march with long strides towards the enemy.' In Frederick's view, 'battalion commanders should push the enemy back, without firing until they have turned to flee.' Frederick was aware of the idea that this sort of advance would be difficult for his men to endure, ordering officers, 'if the soldiers begin to fire, they must reshoulder their muskets and continue to advance, and you may fire by whole battalions as soon as the enemy withdraws.' This was done to preserve the army. Frederick concluded, 'a battle fought in this way will be concluded quite quickly.'[68]

In his desire for a quick battle from a swift advance without firing, Frederick was inspired by his military heroes: Chevalier de Folard, Maurice de Saxe, and Charles XII, all of whom advocated for the advance without firing. Maurice de Saxe made note of Charles XII's use of attacks with *armes blanches*. Charles XII's troops were associated with quick-moving cold-steel attacks, as Johann Maquire noted at Guastalla in 1734.[69] Thus, before the Seven Years War, Frederick was deeply influenced by French and Swedish ideas regarding the importance of attacking with edged weapons. Although he would later retreat from this position, Frederick would provide ammunition for advocates of the *armes blanches* throughout the rest of the century. Adam Storring correctly argues that this was one of many ways that Frederick's art of war looked back to the late-seventeenth and early-eighteenth centuries.[70]

Frederick's comments, and the comments of the authorities he based them on, gave confidence to officers who preferred attacks without firing. Repeatedly, eighteenth-century officers exhorted their men to advance without firing, only using the bayonet in driving the enemy from positions. In 1781, George Washington implored his Continental troops to make this their principal method of attacking the enemy:

> If the Enemy should be tempted to Meet the Army on its March, the General particularly enjoins the troops to place their principle reliance on the Bayonet – that they may prove the Vanity of the Boast which the British make of their particular pains in deciding Battles with that Weapon – He trusts that a generous Emulation will actuate the Allied Armies, that the French whose National Weapon is that of close fight; and the troops in General that have so often used it with success will distinguish themselves on every Occasion that offers – the Justice of the cause in which we are engaged and the Honor of the two Nations must inspire every breast with sentiments that are the presage of Victory.[71]

---

68 Oeuvres de Frédéric la Grand, *Les Principes Géneraux de la Guerre* (Berlin: Imprimerie Royale, 1856), vol.28, p.88.
69 Historical Manuscripts Commission, *Report on Manuscripts in Various Collections*, vol.8, p.408.
70 Adam Storring, '"Le Siècle de Louis XIV": Frederick the Great and French Ways of War', p.19. Accessed at: <https://www.repository.cam.ac.uk/items/55b97a8f-633a-461a-81e2-77dd8a5f2cd0/full>. See also, Adam Storring, *Frederick the Great and the Meanings of War*, unpublished doctoral thesis, St. John's College Cambridge, pp.169–176.
71 LOC: Washington Papers, General Orders, 27 September 1781.

John Burgoyne, in his general orders after the Battle of Freeman's Farm, noted, 'the mistake that [the British troops] are under, in preferring [their fire] to the Baynotte, is to be much lamented. The Lieut. General is persuaded this error will be corrected in the next engagement.'[72] Burgoyne was deeply disappointed, both in his troop's application of firepower, as seen above, and in the lack of reliance on the bayonet.

The Marquis de Silva, a forward-thinking Sardinian officer who supported the use of integrated skirmishers as a 'curtain' to protect battalions in the 1770s, nevertheless argued that in advancing on the enemy a battalion should never fire: 'I do not mean that battalions should never [fire]… but it is indisputable that they should never fire during the charge.'[73] In his view, firing would cause troops that might otherwise close with the enemy to arrest their advance.

In his 1780 tactical manual, Thomas Simes was likewise frustrated with troops seemed unwillingness to close with the enemy. He described the current state of affairs saying, 'and this is what is usually called a charge. It is inconsistent indeed that they should not be able to make a better'. Fortunately for his reader, Simes had the answer: Charges as they presently existed were lackluster 'without the use of the cadence.' Simes continued, 'if the last war had continued some time longer, the close fight [melee] would certainly have become the common method of engaging.' Simes believed this because, in his view, 'the insignificancy of small arms began to be discovered, which make more noise than they do execution.' For him, a reliance of firepower was a crutch, since these weapons 'occasion the defeat of those who depend on them too much.' Therefore, in Simes's mind, 'If… the firings had been laid aside,' how could a battalion formed three or four deep succeed 'against an opposite one, four, or eight deep' marching in cadence, and able 'to perform every movement with more ease'? Simes, like Folard before him, had decided that only firearms stood in the way of fully readopting methods from the classical world, and indeed, the chapter immediately following this forecasting on firearms is entitled 'The Roman Discipline, the cause of their Greatness'.[74]

The comments of an American officer from Maryland, Otho Holland Williams, display how deeply this fascination of advancing without firing ran. Describing the crisis of the Battle of Eutaw Springs in 1781, Williams asserted:

> …the little remnant of the Maryland line, with an intrepidity which was particularly noticed by our gallant commander, advanced in good order, with trailed arms and without regarding or returning the enemy's fire, charged and broke their best troops. Then indeed we fired, and followed them into their camp… the conflict of bayonets decided the victory.[75]

---

72 Edward O'Callaghan, *Orderly Book of Lieut. Gen. John Burgoyne* (Albany: Munsell, 1860), p.116.
73 Marquis de Silva, *Pensées sur la tactique, et la strategique* (Turin: L'imprimerie Royale, 1778), pp.57, 62.
74 Thomas Simes, *A Treaty of Military Science* (London: John Millan, 1780), pp.251–252.
75 Maryland Historical Society: MS 908, Item 116, Otho Holland Williams to Major Edward Giles, 23 September 1781.

Their commander, Nathanael Greene, noted of the battle: 'The gallantry of the officers and bravery of the troops would do honor even to the arms of his Prussian Majesty.'[76] Williams's description matches almost identically with Frederick's prescriptions in *Principes Géneraux*. The troops advanced without firing, and only after the enemy broke did the troops open fire. In noting the prominent role played by the Marylanders, Williams also briefly notes that successful waves of American militia and continentals had previously weakened the British lines with fire before the Maryland advance.

As eighteenth-century officers tried to make sense of their tactical inheritance from the Greeks and Romans, they struggled to find a proper place for the use of firearms. A fantasy that firearms were essentially a passing phase in the history of warfare meant that these men could more freely live in a world where they had more tactical control over their men, in order to better reenact the classical past. This led many of the officers of military Europe to devalue the role and effectiveness of firearms in their own time. The seemingly indecisive nature of musketry firefights gave them hope in this regard: perhaps firearms were less deadly than they seemed.

Conclusion

Between 1733 and 1783, officers across military Europe became infatuated with the Prussian military system. The promise of bringing their own men more tightly under control was an opportunity many of them could not pass up imagining. As a result, after 1763 Prussian military fashion dominated Europe, even among armies, like the British, who remained tactically distinct. Georg von Berenhorst remembered these times in Prussia with a mix of nostalgia and annoyance. Calling Prussomania an 'influenza', he reminded his reader, 'anyone not fortunate enough to attend [the Prussian manoeuvres] was treated to a *"mais je l'ai vu en Prusse."*'[77] In the realm of tactics, officers who hoped to more tightly control their men argued that marching slowly was acceptable if it preserved order, that firepower should be preserved until the last possible moment, and that charging without firing was more decisive than giving fire. These ideas met with the realities of the eighteenth-century battlefield in a process of negotiated authority.

The end result was that the experiential lessons of battle from northern Italy in 1734 were obscured by the dream of a Prussian military machine that could be reproduced in other settings. In the reality of battle beyond the drill square, of course, Prussian soldiers behaved much like their counterparts in other European armies. The next chapter will demonstrate this by examining the negotiated authority which drove most eighteenth-century armies, as well as the sources of royal and religious motivation that drove the Prussians, in particular, to acts of valour.

---

76 Nathanael Green to George Washington, 31 May 1779, in Nathanael Greene and Richard Showman, *The Papers of General Nathanael Greene* (Chapel Hill: University of North Carolina Press, 1986), vol.9, p.360.
77 George von Berenhorst, *Brachtungen über die Kriegskunst* (Leipzig: Gerhard Fleischer, 1827), vol.3, p.289 ('But I saw it like this in Prussia')

5

Negotiated Authority: Inspiration, Religion, and Motivation

The Prussian soldiers were neither machines nor brute beasts.[1]

Christopher Duffy

On 6 May 1787, Prince Henry of Prussia, the younger brother of Frederick II, hosted a lunch for a specific group of veterans of the Seven Years War. The date was no random choice. Thirty years before, at the Battle of Prague, Prince Henry desperately engaged the enemy, leading a flank attack across the Rocketnitzer Bach northeast of the city of Prague. Upon plunging into the stream, Henry had nearly been swept away, until musketeers of the Itzenplitz Regiment waded in after him and bore the prince across the stream on their shoulders. A contemporary tract described the scene:

> Here, this great prince gave proof of his personal bravery. After the Itzenplitz Regiment broke through the enemy line, it had to cross a deep stream. They started to file over, wading in ones and twos. Prince Henry saw this, and leapt from his horse, jumped into the stream and called to them: 'Lads, follow me!' They crossed over and the enemy continued to fall back.[2]

Another account of the incident, perhaps slightly more honest, noted, 'the soldiers stopped at the watery ditch, which seemed deep. Prince Henry threw himself in at once. The smallness of his person increased the magnitude of his potential sacrifice and example.'[3] After the battle, another of the royal brothers, August Wilhelm, recorded, 'My brother [Henry] did wonders. The officers admire him, and the common soldiers swear by him. Heaven be praised that he was preserved, it is a miracle.'[4] Henry would go on to find military success in the Seven Years War, but did not forget the special services

---

1 Duffy, *The Army of Frederick the Great* (2020), p.85.
2 Österreichische Nationalbibliothek: Pk 523, 10 POR MAG, Karl Friedrich Hampe, Schlacht bei Prag 1757.
3 Dietrich Heinrich von Bülow, *Prinz Heinrich von Preussen: Kritische Geschichte seiner Feldzüge* (Berlin: Himburg, 1805), vol.1, p.11.
4 Ernst Berner and Gustav Berthold Volz, *Aus Der Zeit Des Siebenjährigen Krieges: Tagebuchblätter Und Briefe Der Prinzessin Heinrich und des Königlichen Hauses* (Berlin: Duncker, 1908), p.297.

rendered by the Itzenplitz Regiment at Prague. Thirty years later, on the anniversary of the battle, the prince hosted the few surviving officers and men from the Itzenplitz Regiment who had fought with him on that day. Approximately 80 men, including around 70 old common soldiers, turned out to be honoured by the elderly prince.[5]

This episode is a perfect example of the negotiated authority which drove eighteenth-century battles. The soldiers, coming up against what they believed was a perfectly reasonable obstacle, faltered. Officers, or in this case a royal figure, then had to display leadership to overcome the obstacle, rather than simply ordering the men to do so. These Prussians soldiers were not machines: if they were the robotic automata of some officers' desires, they would not have stopped at the ditch. Instead, in the face of obstacles, enemy fire, or approaching danger, battalions might take the perfectly understandable collective decision to hesitate. It required positive leadership on the part of an officer to inspirationally set the attack back in motion. This negotiation between hesitant soldiers and inspirational officers lay at the heart of eighteenth-century infantry warfare.

The concept of negotiated authority in an eighteenth-century military context was first explored in a judicial setting by William P. Tatum III.[6] Tatum usefully links the idea of a negotiated authority and moral economy to the military justice system of the eighteenth-century British Army. Seanegan P. Sculley has utilized the concept of negotiated authority in the American War of Independence, but assumes it applied mainly to the American Continental Army.[7] In the context of the American War of Independence, Stephen Conway has examined the idea in the social life of armies.[8] Conway helpfully defines negotiated authority:

> …the give-and-take that it implies, the need for the nominal superior to secure the agreement or at least acquiescence of the nominal inferior, makes it applicable to the way in which eighteenth-century army officers were obliged to operate. Their authority was far from absolute; they could not take it for granted that their men would always obey. Only by keeping within the lines drawn by the military moral economy and their soldiers' contractual attitudes could officers effectively run their units.[9]

---

5   Franz Ludwig von Haller, *Militärischer Charakter und merkwürdige Kriegsthaten Friedrich des Einzigen: Königs von Preussen: nebst einem anhang über einige seiner berühmtesten Feldherren und verschiedene preussische Regimenter* (Berlin: Jüngern, 1796), p.320.
6   William p.Tatum III, '"The Soldiers Murmured much on Account of this Usage": Military Justice and Negotiated Authority in the Eighteenth-Century British Army', in Kevin Linch and Matthew McCormack, *Britian's Soldiers: Rethinking War and Society, 1715–1815* (Liverpool: Liverpool University Press, 2014), pp.95–113.
7   Seanegan P. Sculley, *Military Leadership in the Continental Army* (Yardley: Westholme Publishing, 2019), pp.vii–xxxiv.
8   Stephen Conway, 'Moral Economy, Contract, and Negotiated Authority in American, British, and German Militaries, ca. 1740-1783', *The Journal of Modern History*, vol.88, no.1 (March 2016), pp.34–59.
9   Conway, 'Moral Economy, Contact, and Negotiated Authority', p.35.

Both Tatum and Conway applied this concept to the majority of army life – that is to say, outside combat. Conway largely focuses on issues such as soldiers insisting on regular payment, avoiding brutal work details, ensuring a regular supply of food, and resolving mutinies over poor treatment. There is no analysis of how the idea of a moral economy or negotiated authority might be applied to the battlefield.

Within the world of combat, negotiated authority forms one way to understand when soldiers lost motivation: why some battalions ran, and others stood in place. Why did eighteenth-century soldiers turn and flee? Katrin and Sascha Möbius have provided the fullest exploration of Prussian soldiers' motivations, and Ilya Berkovich has examined the motivations of eighteenth-century armies as a whole.[10] Without entering into the debate over masculinity in combat, whether music represented 'the fine tuning of military honour,' or whether or not the 'tactical body' was a 'space of honour,' suffice it to say, Ilya Berkovich correctly asserts that 'fear of dishonour formed the single most important collective source of combat motivation between 1700 and 1800.'[11] Samuel Dodson has also helpfully shown that eighteenth-century officers wrote and thought about their soldiers' emotions in combat a great deal.[12]

Thus, officers felt the need to provide direction on the battlefield, not just by forcing men to adhere to their tactical doctrines, but by leading them in an inspirational way. This chapter is the first exploration of the negotiated authority which marked the relationship between officers and men in combat.

Inspirational Officers and Negotiated Authority

The French *lieutenant général*, Jacques Marie Ray de Saint-Geniez, reflecting on his military service between 1741 and 1762, was reminded of the successes of his opponent, Karl Wilhelm Ferdinand, the Hereditary Prince (*Erbprinz*) of Brunswick. The chevalier de Ray recalled, 'The Hereditary Prince often lacks that certainty of judgement which allows one to decide to abandon or pursue an attack against an enemy position.'[13] The French officer swiftly turned this insult into a backhanded compliment, by noting, 'It was the enthusiasm that he aroused in his soldiers; the affection of his men, and the confidence that it inspired, which allowed this prince… to oppose the operations of our armies.'[14] Indeed, the chevalier de Ray recalled that the Prince had asserted:

'It is in the nature of soldiers to never abandon officers that they esteem and love. We must do the same for them. Visit them in their tents, move to

---

10   Berkovich, *Motivation in War*; Katrin and Sascha Möbius, *The Psychology of Honour*.
11   Ilya Berkovich, 'Fear, Honour and Emotional Control on the Eighteenth-Century Battlefield,' in Erika Kuijpers and Cornelis van der Haven, *Battlefield Emotions 1500-1800: Practices, Experience, Imagination* (Basingstoke: Palgrave Macmillan, 2016), pp.93–110.
12   Samuel Dodson, 'Courage, Honour, and Phlegm: A Study of Eighteenth-Century Military Writers' Descriptions of Soldiers' Combat Emotions and Motivation', *Journal of Eighteenth-Century Studies*, vol.46, no.2, (2023), pp.279–295.
13   Jacques Marie Ray de Saint-Geniez and Lucien Mouillard (ed.), *Réflexions et souvenirs du chevalier de Ray* (Paris: Charles-Lavuzelle, 1895), p.231.
14   Ray and Mouillard, *Réflexions et souvenirs*, p.231.

where the advanced guards are fighting and only part company when you die beside them. With these sentiments officers will train their soldiers well and become distinguished in the actions of war.' So said the Hereditary Prince, and I agree with him completely.[15]

Ray's quoted sentiment from the *Erbprinz*, whether real or not, contains the key phrase, *Il faut leur render le pareille*, (lit. 'we must give them the same,' rendered above 'we must do the same for them.'), which expresses the contract at the core of negotiated authority. Soldiers had a duty to be brave and loyal to their officers, and never to abandon them. The officers, in their turn, had the duty to provide leadership, paternal affection, and love to their men. Thus, as Prussian officers went into action, Austrian reports noted that they called out to their men: 'Lads, behave like the brave soldiers you are: we won't abandon you, and don't you abandon us!'[16] This contract – to fight and die together – bound officers and men of the same battalion together in combat. It formed a key component of the dishonour that flight from combat entailed. Ray's sentiment – *et ne s'en séparer qu'à la mort à côté d'eux* (lit. 'and only part with them when you die beside them') – likewise sounds similar to a traditional phrase: *jusqu'à la mort nous sépare* (till death do us part). Soldiers were not quite married to their officers, but these two groups did share a trust to not abandon the other. As a result, eighteenth-century officers often led from the front, to uphold their share of the bargain.

An officer who almost perfectly lived out *Erbprinz*'s maxim was *Hauptmann* Ludwig Heinrich Vollrath von Erckert of the Ansbach-Bayreuth Grenadiers in the American War of Independence. During the assault on the American-held Fort Montgomery in October of 1777, von Erckert was badly wounded in the right arm by a cannister ball. Despite losing the use of his arm, he picked up his sword with his left hand and called to his grenadier company: 'Be consoled and undisturbed, my children! I still lead you bravely on and will not leave you. Only press forward! Keep your honour, keep your courage!'[17] Von Erckert was mortally wounded by another blast of cannister shortly afterwards, but unlike most situations of this type, his men continued the assault, taking the American position. The men followed his instructions, knowing that their captain had completely fulfilled the unspoken martial contract between officers and enlisted men in combat. Under the military codes of the eighteenth century, it would have been perfectly acceptable for von Erckert to leave the field to seek medical treatment. His failure to do so, and his continued leadership after his first wounding, was highly inspirational to his men.

Hesitation was a normal part of fighting in varied terrain: with the enemy firing on you, crossing a stream, wading through snow, and charging uphill into enemy defences were terrifying. Officers understood this, but also knew their duty was to keep the battalion moving forward to drive the enemy from the field.

---

15   Ray and Mouillard, *Réflexions et souvenirs*, pp.231–232.
16   Quoted in Duffy, *The Army of Frederick the Great* (2020), p.103.
17   Johann Conrad Döhla and Bruce Burgoyne (ed.), *A Hessian Diary of the American Revolution* (Norman: University of Oklahoma Press, 1990), pp.52–53.

General Leutnant Prinz Moritz von Anhalt-Dessau commanded our left wing, consisting of the Grenadier Battalion Schoening and the four infantry regiments Prince of Prussia, Prinz Dietrich, Prinz Leopold, and Bredew, of the first line, came to a difficult snow-filled ditch, it seemed impossible for the infantry to pass. This intrepid Prinz led the way together with two musketeers from the Prinz Dietrich Regiment. They sprang into the ditch, passed through, and the whole brigade followed in the same manner.[18]

Having passed their way through the ditch, Prinz Moritz then led the Prince of Prussia Regiment forward charge, sword in hand, to drive off the enemy infantry opposing them. Having shown that he was willing to lead from the front in a moment of hesitation, Prinz Moritz had earned his men's trust as he led them to the attack. Motivating men to conduct bayonet charges at close range often required just this sort of intervention, particularly if the troops charging had previously fired.

In this sorry state of affairs, the Duke of Bevern [August Wilhelm, Herzog von Braunschweig-Bevern] came galloping up, passing through heavy fire like an intrepid hero, and saw how the lads of his regiment, because of the impractical terrain, were not in close order. Rather, they had to fight in small groups on the hillside, and did not reply to the enemy's heavy volleys. 'Children,' yelled the Duke, 'Shoot for God's sake, shoot and advance!' 'Oh, dear father,' the lads replied, 'What shall we do?' We have no more powder, and are being shot dead without reply!' 'What?' cried the Duke, 'Don't you have bayonets? Go and kill the dogs!'[19]

Here, we see one of the celebrated commanders of Frederick's Prussia conversing with common soldiers of his own regiment. As the *inhaber* of the unit, Bevern often had other responsibilities, but had a contractual relationship to look out for the welfare of his men. Thus, the familial and paternal language (*'Kinder/lieber Vater'*) used by both parties in this exchange had special significance, perhaps even more than its general use by officers and men. Bevern's success in this attack was recorded by more than one source. This account, rather flattering to Bevern, was recorded by a source close to him: his secretary. As a result, it is helpful that we have an additional account of the same action from another officer: 'The men complained that they had no ammunition. The Duke called out, "Lads, don't be frightened! Otherwise, why would we have taught you to use your bayonets?"'[20] Although the familial language is absent from this account, it still has Bevern employing the diminutive *Bursche* (lit. lads/boys). The second source, the officer Retzow, described the effect: 'These words from the mouth of a leader in which the soldiers had complete faith, were like God's

---

18 Anon., *Sammlung ungedruckter Nachrichten so die Geschichte der Feldzüge der Preußen von 1740 bis 1779* (Dresden: Walther, 1785), vol.1, pp.430–431.
19 Curt Jany, *Urkundliche Beiträge und Forschungen zur Geschichte des preussischen Heeres* (Berlin: E.S. Mittler, 1901), vol.1, pp.9–10
20 Friedrich August von Retzow, *Charakteristik der wichtigsten Ereignisse des siebenjährigen Krieges* (Berlin: Himburg, 1802), vol.1, p.63.

own word for the Prussians. They immediately closed ranks, and with the example of their officers, threw themselves at the enemy.'[21]

So far, all of the sources utilized have come from officers and their literary colleagues, but what did the soldiers themselves think of these inspirational examples? British Sergeant Roger Lamb noted the inspiration intervention of Lieutenant Colonel James Webster in a moment of British hesitation. The outnumbered British forces were preparing to launch an attack on the American first line at the Battle of Guilford Courthouse. The Americans were prepared along a rail fence, taking careful aim at the British. This sight gave the British pause:

> At this awful period a general pause took place; both parties surveyed each other for the moment with the most anxious suspense. Nothing speaks *the general* more than seizing on these decisive moments: colonel Webster rode forward in front of the 23rd Regiment, and said, with more than even his usual commanding voice… *'Come on, my brave Fuzileers'* This operated like an inspiring voice, they rushed forward amidst the enemy's fire; dreadful was the havoc on both sides.[22]

In this passage Lamb asserts that from the perspective of an enlisted man, nothing was more impressive than an officer taking initiative in a moment of hesitation. For Lamb, it was the highest form of leadership, the very essence of being a commander. Webster was mortally wounded in the attack, but the British attack continued, eventually driving the Americans from the field at a heavy cost. The intervention from an officer could have an equally powerful effect on men who were retreating or wavering, as another enlisted man recalled.

Sergeant Nathaniel Root was having a very poor day at the Battle of Princeton, he had been wounded, his platoon had been shot to pieces, and he was in the process of seeking safety. Then, General George Washington intervened:

> [I] discharged my musket at a part of the enemy, and ran for a piece of wood, at a little distance where I thought I might shelter. At this moment, Washington appeared in front of the American Army, riding towards those of us who were retreating, and exclaimed, 'Parade with us, my brave fellows, there is but a handful of the enemy, and we will have them directly.' I immediately joined the main body and marched over the ground again.[23]

Although Root was less effusive of his praise for Washington than Lamb was for Webster, he notes the immediate effect. His retreat was stopped, and he returned to the attack at the place where his unit had been mauled and he had been wounded. So far we have seen a litany of successes: both European and American examples

---

21 Retzow, *Charakteristik*, vol.1, pp.63
22 Roger Lamb, *An Original and Authentic Journal of Occurrences during the Late American War from its Commencement to the Year 1783* (Dublin: Wilkinson & Courtney, 1809), p.361.
23 Nathaniel Root, 'The Battle of Princeton,' *Pennsylvania Magazine of History and Biography*, vol.20 (1896), p.517.

of officers decisively intervening to good effect. It is equally important to note that a misstep, wounding, or death in the moment of an inspirational intervention could have disastrous effects. We have already seen the case of the Piedmontese officer Lucadou and the Austrian grenadier officer at Parma, but the case is worth repeating. Here, Warnery describes the effects of seeing an officer killed in the moment of leadership.

> As the grenadiers were hesitant to pass the ditch, quite deep, one of the officers, no doubt a fire-eater, descended, and went back up in order to urge them to follow his example, with no other weapon but the sword in his hand… Lucadou, an intrepid man, although not five feet tall, approached him and giving him a bayonet thrust, at the same time shot him through the body and he fell. The fall of the German dismayed his soldiers so much that they no longer thought of crossing the ditch, not even to retrieve his body. I could cite other examples where an officer was killed, and his troops were completely dismayed.[24]

The Austrian grenadier officer had tried to inspire his men, failed in the face of the enemy, and was abandoned by his men after his death. This death, however, garnered the respect of Warnery, who even after 40 years noted, 'I am sorry I do not have the name of this brave man.'[25] Warnery also used this case as a powerful rebuttal to theorists who claimed officers did not need to carry fusils. Lucadou had apparently admitted to Warnery at a later date that he only attempted to kill the Austrian giant because he had a bayonet and fusil, and the grenadier was only armed with a sword. When leading from the front, it paid to bring the right tools for the job. For all of its intricacies, this story shows marked similarities to Prince Henry at the Rocketnitzer Bach, and Prinz Moritz at Kesselsdorf. When trying to cross obstacles, particularly when under enemy fire, soldiers hesitated and often needed someone to urge them along.

Of course, no discussion of inspirational leadership in the mid-eighteenth century would be complete without its most famous heroic death: Prussian *Feldmarschall* Kurt Christoph von Schwerin at Prague in 1757. Like Prince Henry's dive into the stream, Prussian printmakers loved the image of Schwerin's final charge. What are the specifics of this incident, though? Once again, Warnery described the scene:

> We know that the Prussian commanders pride themselves on having [been the *inhaber* of] good regiments…The Marshall [Schwerin] was really proud of his regiment, which was always distinguished, and that he had cultivated. As the attack started he had his eye on it, and frustrated to see it going into disorder, he rushed to it, took a flag in his hand, and shouted, 'Whoever is brave, follow me!' Whereupon, many of the officers and most of the soldiers did. He was then killed by a blast of canister, and his regiment

---

24 Charles de Warnery, *Anecdotes et pensées historiques et militaires, écrites vers l'année 1774* (Halle: Jean Jaques Court, 1781), p.17.
25 Warnery, *Anecdotes et pensées*, p.17.

was repulsed so badly that Winterfeldt… had real difficulty in stopping their flight, shouting after them.[26]

In seizing the colours and leading his men forward in a desperate moment, Schwerin was not just an inspirational army commander; he was also being a paternal figure to *his* regiment. As the *inhaber*, he had a particular duty to care for his men and, in this case, share their risks. As it turned out, Schwerin's decision had fatal consequences for himself and the Prussian attack, but he was following an established pattern of leadership as a Prussian officer, an *inhaber*, and a gentleman of military Europe. When the soldiers wavered, or the attack began to falter, it was the duty of an officer to attempt to personally restore the situation and set the men back in motion forward.

At Paltzig in 1759, *Preimerleutnant* Jakob Friedrich von Lemcke found himself in a similar desperate situation. His unit was stationed on the edge of a wood, looking out at Russian cannon deployed on the plain in front of them. Lemcke believed that his moment of glory had arrived and desperately wanted to try and take the cannons:

> I pointed out the [Russian] cannons to General Hülsen, who was riding just behind my platoon, and asked him to help me so that together we could take the guns. He told me that I should stick with my men as long as I could, and he would try to order up cavalry to support me. I therefore ordered my men to advance, but they had no intention of advancing at all, but took cover behind trees and fired at the enemy from there. I had already bent my sword beating my men, which I held instead of my spontoon, but in no way could I get them to advance. A cannonball then came bouncing through and smashed my smashed my left foot, and I fell to the ground. My men immediately fled and I was totally deserted.[27]

Here, we see the inverse of the positive leadership so often displayed in military Europe. When officers resorted to physical punishment, as opposed to inspirational words and deeds, it rarely had the desired effect. Lemcke's troops – a platoon of infantry from the Anhalt-Dessau Regiment – were quite content to fire at the artillerymen from cover; they had no desire to face down a blast of cannister in the open. Lemcke responded to this situation, not by trying to inspire his men, but by beating his men for their failure to follow his orders. In doing so, he eroded any sort of trust which might have existed between them. When the first opportunity arose, they deserted him to Russian captivity.

These collected examples demonstrate much regarding negotiated authority on the battlefield. Troops responded well to a more positive leadership when being asked to advance in the face of enemy fire. This was by no means always successful,

---

26 Charles Emmanuel de Warnery, *Campagnes de Frédéric II, Roi de Prusse, de 1756 à 1762* (Location and publisher not given, 1788), pp.111–112.
27 Jakob Friedrich von Lemcke and R. Walz (ed.) 'Kriegs-und Friedensbilder aus den Jahren 1754–1759', *Preußische Jahrbücher*, vol.138 (1909), p.36.

and frequently the officer who attempted an inspiring act fulfilled chevalier de Ray's vision of dying by the side of his advanced guards. In some of these circumstances, the attempted inspirational act backfired, sending the men into retreat, and any chance at a quick victory.

Despite the risk, however, these inspirational acts were the currency of negotiated authority for officers. Simply beating the men who refused to follow orders had little effect, as Lemcke found in 1759 and Johann Maquire could have told him 25 years earlier at Guastalla. Officers who simply opined that the men would have won 'if they had followed the orders of the commander' and beat their troops rarely found success.[28] By contrast, exposing yourself to enemy fire and offering to lead your men forward carried significant personal risk, but gave the officers a solid position in the negotiation of authority. Across military Europe, soldiers also turned to other sources of inspiration to motivate them in combat: they were inspired by their places as servants of the King, and perhaps most importantly, the religious faith which sustained them.

God and the King: Monarchical and Religious Inspiration

The remainder of this chapter picks up an important theme from the last: what made the Prussian army so exceptionally successful in the middle decades of the eighteenth century? Military observers hoped that they could unlock the secret to Prussian success by copying the externals of uniforms and drill manuals. Only the most astute observers noted perhaps Prussian success did not derive from its rigid drill, but from the motivations of its common soldiers.

In Prussia, displays of religious feeling among common soldiers were communal and army wide. Sascha and Katrin Möbius have examined religion in the Prussian Army extensively, as a combat motivation, exploring the use of prayer and calls for prayer and intercession, and a belief in the providence of God.[29] This chapter confirms and enhances their arguments, and using letters as a source base, explores the attributes that Prussian soldiers developed in their view of God, not just in combat, but as they experienced life in the Prussian army and beyond.

As a result of the agreements reached between King Frederick William I and the Halle Pietist movement in the 1710s, Pietists provided chaplains, or *Feldprediger,* for the Prussian military. Although common soldiers initially viewed these chaplains with suspicion, by the 1740s the Pietists had made serious inroads into the Prussian army. During battle, *Feldprediger* often stood alongside their soldiers, leading them in the singing of hymns and attempting to inspire them. Pastor Joachim Friedrich Seegebarth rallied the fleeing regiment of Erbprinz Leopold at the Battle of Chotusitz, he recalled:

---

28   Historical Manuscripts Commission, *Report on Manuscripts in Various Collection*, vol.8, p.408.
29   Sascha Möbius, *Mehr Angst vor dem Offizier als vor dem Feind,* pp.107–111, Katrin and Sascha Möbius, *Prussian Army Soldiers and the Seven Years War: The Psychology of Honour,* pp.90–92, 135–137.

> Our regiment was falling back and became partly mixed with the enemy cavalry and grenadiers. I came up, and called out the soldiers and officers in moving and serious terms that they must stand fast and rally. A number of them were willing enough and at once answered with a loud, 'yes!'... After my intervention the bullets flew about my head like a swarm of buzzing gnats, but thanks be to God my coat was not even touched.[30]

In addition to this battlefield role, chaplains recorded notes about individual soldiers in their record books, including soldiers promising to live more holy lives.[31] The chaplains played an active role in soldiers' lives; they prayed with soldiers in the morning and evening and served as the teachers for soldiers' children during the working day.[32]

By the 1750s, Prussian common soldiers frequently employed religious language in their writings. *Feldwebel* G.S. Liebler, a Pietist from the Halle region who served alongside his equally religious son, J.S. Liebler, noted soldiers' devoted attendance at daily prayer meetings:

> I rejoice in God whenever I attend our daily prayer meetings, which are always held at 10 a.m. by the whole army. I even see devotion and awe on the part of our superiors, though not in all, and this spiritual leadership awakens some of the soldiers. Lord Jesus, awaken them all, that all of us may become spiritual champions for your honour and for the benefit of the Christian Church.[33]

Likewise, some Prussian soldiers had adopted the Lutheran language which framed human experience as a struggle between two kingdoms: the Kingdom of God and the Kingdom of this world. Contemporaries wrote regarding the 'Children of God' (*Kinder Gottes*) and 'Children of the World' (*Weltkinder*).[34] At least some Prussian soldiers framed their existence along similar terms, but employed the idea of Satan and the World, or Worldly, (*Satan die Welt, Weltlich*) as opposed to *Weltkinder*.[35] Devout soldiers viewed themselves as the *Kinder Gottes*, and admonished their worldly comrades as *Weltkinder*.

Most Prussian soldiers emphasized two aspects of God in their writings: God's love and his power. These two aspects of God fit well together in the worldview of Prussian soldiers. A loving God cared for them personally and reflected the love that

---

30 Georg Heinrich Von Berenhorst and Eduard von Bülow, *Aus Dem Nachlasse Von Georg Heinrich Von Berenhorst* (Dessau: K. Aue, 1845), pp.99–100.
31 Carl Hinrichs, *Preussentum und Pietismus: Der Pietismus in Brandenburg-Preussen als Religiössozial Reformbewegung* (Göttingen: Vandenhoeck & Ruprecht, 1971), p.164.
32 For more on the Pietist Chaplaincy, see, Benjamin Marschke, *Absolutely Pietist: Patronage, Factionalism, and State-Building in the Early Eighteenth-Century Prussian Army Chaplaincy* (Leipzig: Max Niemeyer Verlag, 2005).
33 Curt Jany, *Urkundliche Beiträge Und Forschungen Zur Geschichte des Preussischen Heeres* (Berlin: E.S. Mittler, 1901), p.34.
34 Jany, *Urkundliche Beiträge*, pp.55, 42.
35 Jany, *Urkundliche Beiträge*, p.42.

they felt for their families and their homes. A powerful God could provide personal protection against sickness and the horrors of the battlefield, and also ensure the overall victory of the Prussian Army against the many enemies it faced. Likewise, a loving and protecting God would understand soldiers' fears for their families and take those loved ones into his protective care during their absence. Barthel Linck wrote briefly to his wife after the Battle of Lobositz. His words sum up the Prussian soldier's view of God: 'The merciful, benevolent, faithful, and loving God has fulfilled this powerful promise of divine protection for our army, showering us with undeserved grace and worth today, on the 1st of October, out of his fatherly divine love.'[36]

Prussian soldiers emphasized God's love for them as individual soldiers as well as protecting their king and cause. Johann Hermann Dresel wrote of the divinely 'dear' or loving (*Liebe*) God's ability to grant peace that would end the war as well as the knowledge that such an affectionate God possessed.[37] Barthel Linck noted that 'God wants to take us into Grace,' and although he could not write all the details in his letter, he would tell his wife about everything 'when the loving God brings me back to you.'[38] Herr Kistenmacher, the secretary of August Wilhelm, Duke of Braunschweig-Bevern, noted, 'The loving God has protected the Duke, though he stood in the strongest fire.'[39] Frantz Reiß asked his wife to 'join your prayers with mine, that the loving God will protect me further.'[40] Many Prussian soldiers referenced God's love in their writing and united the idea of his love for them with the concept of divine protection.

In the worldview of Prussian soldiers, God was an all-powerful being who took a direct interest, not only in the outcome of particular battles, but the survival of individual soldiers. In essence, God was the ultimate patron who could be called upon to give their army victory, provide them with protection, and safeguard their families during their absence. Prussian soldiers frequently noted that without God's assistance they would have been defeated. Kaspar Kalberlah asked his family to pray 'God does not allow this enemy to become too mighty and conquer us, otherwise we will all fare badly.'[41] Frantz Reiß noted that the success at the Battle of Lobositz would have been impossible without divine protection: 'But alone, our small group would have been insignificant against [the Austrians], if God had not been on our side we all would have been struck down, but thank God, it is finished.'[42] In the same vein, soldier J.S. Liebler instructed his family to 'praise the goodness of the Lord, for He has done great things for us.'[43] Using the exact same phrase in a letter over six months later, his father *Feldwebel* Liebler noted before the Battle of Prague, 'The Lord has done great things for us, if the Lord were not with us, our enemies would

---

36 Jany, *Urkundliche Beiträge*, p.12.
37 Kohl, 'Ein Brief des Wiblingwerder Bauernshones Johann Hermann Dresel, pp.82–83.
38 Jany, *Urkundliche Beiträge*, p.15.
39 Jany, *Urkundliche Beiträge*, p.10.
40 Jany, *Urkundliche Beiträge*, p.32.
41 Liebe, *Preussische Soldatenbriefe*, pp.29–30.
42 Jany, *Urkundliche Beiträge*, p.31.
43 Jany, *Urkundliche Beiträge*, p.16.

have devoured us.'⁴⁴ *Unteroffizer* Müller wrote to his wife after the Battle of Prague: 'I am still in good health, God be praised, and believe that our actions will continue to go well, with divine assistance.'⁴⁵ Müller linked two ideas: God's care for the army as a whole and care for his health personally. Even as Prussian soldiers trusted that God would ensure victory over the enemies of their Prince and state, they entrusted their personal well-being to Him.

Soldiers believed that God directed not just the flow of battles as a whole, but the individual bullets which could harm them. Katrin and Sascha Möbius have noted the way in which this psychologically boosted soldiers in combat and have demonstrated that religious thoughts of personal protection helped keep soldiers from flight.⁴⁶ Soldiers linked survival in combat with divine protection and seem to have especially cherished close calls with death as evidence that God had miraculously preserved them. Johann Jacob Dominicus noted in a letter to his brother that God had taken special care of him at the Battle of Paltzig, and recorded this in some detail:

> I have four signs that show how miraculously the Lord saved me: one bullet went through the tip of my hat, one through the turnback, one went into the butt of my musket and a part of the cover of my cartridge pouch was shot off. While I was standing and loading, a bullet flew just over my hand and bent my ramrod like a fiddlestick. Next to me on my left, the legs of three men were evenly shot off at the same height – there is no doubt that this had been done with chain-shot. The man to my right was wounded and the one next to him killed… But I saw that my time had not yet come, and was so full of sorrow that I could not give thanks to God nor sing a song of praise. Help me with that, my friends! To thank God that he protected me and ask him to further safeguard me due to his grace, because it has not come to an end, yet. Many of us defect, but I will not break my oath if God further grants health and life to me, I will remain faithful to God and the king and will bear the burden as long as God wants me to. I have often experienced many temptations and tribulations, but the Lord God has maintained my good thoughts and I will lead my life and actions in a way I can answer for to God and man.⁴⁷

Dominicus's statements are nothing short of extraordinary, giving us a window into the visceral and psychologically demanding nature of eighteenth-century combat. Dominicus offers evidence of deadly combat and divine protection but could not bring himself to praise God for it as a result of the shock and horror of near death. Indeed, he requests assistance from his friends and promises to maintain his loyalty to both God and the king.

---

44   Jany, *Urkundliche Beiträge*, p.40.
45   Jany, *Urkundliche Beiträge*, p.58.
46   Möbius, *Prussian Army Soldiers*, p.136.
47   Dominicus, *Aus dem Siebenjährigen Krieg*, pp.62–63; See Katrin and Sascha Möbius, *Prussian Army Soldiers*, pp.178–181, for a full translation of this letter.

Far from being unique, Dominicus's horrific experiences are echoed in the writings of other Prussian soldiers from the era of the Seven Years War. Frantz Reiß recalled the shocking experience of the first major battlefield encounter of the war at Lobositz:

> So the battle began at six o'clock in the morning and dragged on amidst thundering and firing until four in the afternoon, and all the while I stood in such danger that I cannot thank God enough for my health. In the very first cannon shots [my friend] Krumpholtz took a cannon ball through his head and the half of it was blown away, he was standing just beside me, and Bode [was hit by Krumpholtz] brains and pieces of his skull and my musket was blown to pieces from my shoulder, [he] praise God, was uninjured. Now, dear wife, I cannot possibly describe what happened, for the shooting on both sides was so great, that no-one could hear a word of what anyone was saying, and we didn't see and hear just a thousand bullets, but many thousands. But as we got into the afternoon, the enemy took flight and God gave us the victory. And as we came forward into the field, we saw men lying, not just one, but 3 or 4 lying on top of each other, some dead with their heads gone, others short of both legs, or their arms missing, in short, it was a horrifying sight. Now, dear child, just think of how we must have felt, we who had been led meekly to the slaughterhouse without the faintest inkling of what was to come.[48]

In his same letter, Reiß paraphrased the regimental scripture reading which his regiment listened to after the battle.[49] Like Dominicus, he notes the death of soldiers around him – in this case a man who was known to his wife and local community, Krumpholtz. Reiß felt the horror of this loss keenly. *Feldwebel* Liebler also recorded brushes with death as a sign of God's protecting power:

> Now a battalion of grenadiers advanced, whereby General von Ingersleben commanded us to follow, and again we came under the hail of canister shot, and here I learned what it means to be under the shield of the Most High. A canister ball hit me on the sternum, and I certainly thought that my life was at an end, but the bullet did not pierce my flesh, I staggered a few steps back but still stood under the rain of fire, until the grenadiers had to give way under the fire of this battery.[50]

For Liebler, Reiß, and Dominicus, facing heavy fire and having close brushes with death were signs of divine protection. These men, though they saw scores of men cut down around them, believed that their lives had been marked for a special purpose by God. Their continued life, they argued, was a sign that God's protection was as tangible as the muskets they held and the bullets they fired.

---

48 Jany, *Urkundliche Beiträge*, p.30.
49 See: Möbius, *Prussian Army Soldiers*, p.183.
50 Möbius, *Prussian Army Soldiers*, p.47.

Religious sentiments were crucial in Frederick's ability to cultivate loyalty among Prussian natives and foreign troops via mass-produced familiarity. Scripture, songs, and liturgical rituals helped to solidify his authority as king and military commander. *Generalleutnant* von Lossow observed, 'The common soldier regarded the king as the representative of God, whom he must fear, honour, and love.'[51] The letters of Prussian common soldiers, as discussed above, are full of religious language, and as a familial yet stern, harsh but caring figure, Frederick worked his way into the cosmology of his men. Sascha Möbius has asserted that 'God and the King' was the unofficial watchword of the Prussian army, and it seems that linking divine protection with royal authority turned into something of an ideology during the Seven Years War.[52] *Feldprediger* and author Karl Daniel Küster formulated this ideology more fully in a sermon to his troops in 1758, arguing that it had three parts: '1) God Lives. 2) The King Lives – 3) I will be true to God and the King until death.'[53] As a *Feldprediger,* Küster had a bias towards ascribing religious motivation to his men, but it is clear that this phrase was internalized by Prussian soldiers.

This ideology of 'God and the King,' summarized by Küster, can be seen in the writings of Prussian common soldiers. Prussian Musketeer Johann Riemann indicated that this sentiment was alive and well in some of the darker days of the Seven Years War. In the summer of 1762, after the death of his brother and fellow soldier Benjamin, Riemann wrote:

> In such mortal danger, I have long survived, I will last the year if it pleases God, in such hunger as makes life miserable, none of us have heard or experienced it before. We ask God daily to have mercy on us and end the misery of life. We want to trust in God, who has helped us through many a sad time. As our enemies all circle: They are ours. God will assist the Prussian Army, and not forget his promises. God and Frederick still live.[54]

'God and the King' was indeed the watchword of the Prussian army. After the defeat at Paltzig in July 1759, Johann Jacob Dominicus wrote home to his brother: 'I will remain faithful to God and the King and bear this burden as long as God wants me to.'[55] An Anhaltiner in the Prussian army, Hoppe, recalled his sentiments for God and Frederick during the Zorndorf campaign of 1758. Speaking on the experiences of Anhaltiners, Hoppe wrote, 'We foreigners did not grumble at these heavy tasks, because we knew quite well that we could not serve anywhere better than under the King, who saw the hardships of the common man and shared them... I never would have lasted so long without my faith in God and the King.'[56] With a loving and

---

51 Ludwig von Lossow, *Denkwürdigkeiten Zur Charakteristik Der Preussischen Armee Unter Dem Grossen König Friedrich Dem Zweiten* (Glogau: Carl Heymann, 1826), p.3.
52 Möbius, *Leben in der Stadt*, p.92.
53 Küster, *Bruchstucke*, p.74. See also: Ernst von Barsewisch, *Meine Kriegs-Erlebnisse während des Siebenjährigen Krieges, 1757–1763* (Berlin: Warnsdorff, 1863), p.23.
54 Liebe, *Preussische Soldatenbriefe*, pp.34–35.
55 Dominicus, *Aus dem Siebenjährigen Krieg*, p.63.
56 Anon., *Offizier-Lesebuch, Historisch-militärischen Inhalts, mit untermischten Interessanten Anekdoten, Von Einer Gesellschafts Militärischer Freunde* (Berlin: C. Matzdorff's

powerful God on their side, and a king who promised to care for them and listened to their complaints, Prussian soldiers emphasized God and the king in explaining their loyalty, even as it became apparent that Prussia was not winning the wider war.

In the writings of most of his soldiers, Frederick remained a personable yet distant figure. A perfect illustration of this is provided by Christian Friedrich Zander, describing the aftermath of the Battle of Zorndorf in 1758. The men of Zander's regiment saw the king riding behind them, and he shouted:

> Good morning, lads, are you still pretty healthy? Yes, the lads answered, but you should have taken us along to the Russians. He answered: You do not have to be around everywhere. Be patient. You shall soon make money with the Austrians. There you shall get good Kremnitz Ducats as a booty. They are better than rubles.'[57]

This friendly interaction between Frederick and his men was typical in the Seven Years War and can be observed in a number of sources from common soldiers, including letters and diaries written during the war, officer's magazines published after the conflict ended, and interviews with old veterans in the 1820s.[58]

This type of story became a sort of ritual or exchange, almost on the level of a liturgical experience for Prussian common soldiers. The exchange usually began with the king offering a greeting and following up with a leading question, which the body of troops could collectively answer in the affirmative. The men usually followed their affirmative answer with a question regarding military life or a tentative recommendation. The king ended the exchange with an appeal for patience and a positive prediction of future events. Frederick certainly used this method of addressing soldiers in the middle years of the Seven Years War, as many sources describe it between 1757 and 1759. This type of exchange in the Prussian service became so ubiquitous that its use was noted by Austrian governmental reports, which indicated that 'the only way officers talk to soldiers is to say things like, "cheer up lads, it will get better soon!" Words of this sort cost no money, but they encourage the soldiers in an extraordinary way, and establish affection for their officers.'[59] Like the king, Prussian officers drew on a similar paternalistic exchange. Frederick and his officers were able to use this type of interaction to create a sense that they were familiar with the challenges that soldiers faced in army life and interested in creating a moral economy of soldiering.

Buchhandlung, 1793), pp.180–182.
57 'Kremnitz Dukaten' were gold coins produced in the Habsburg mint in Kremnitz (today Kremnica in Slovakia) and valued for their purity. This letter has not survived to the present, but a report of its contents is in: Zander, *Fundstücke*, pp.109–110.
58 Dominicus, *Aus dem Siebenjährigen Krieg*, p.59; C. Hildebrandt, *Anekdoten und Charakterzüge aus dem Leben Friedrichs des Grossen* (Halberstadt: Brüggemann, 1829), vol.2, p.39; Anon., *Offizier-Lesebuch, Historisch-militärischen Inhalts, Mit Untermischten Interessanten Anekdoten*, pp.185–186; Zander, *Fundstücke*, pp.109–110.
59 Kriegsarchiv, Vienna (KV): Kriegsakten 387, Verzeichnis deren vornehmsten und wichtigsten Vortheile.

It is worth pausing for a moment to consider the significance of this ritual. The king could not be familiar with the tens of thousands of men in his army on an individual level. As a result, this exchange provided the king with the ability to seem approachable to soldiers, as it could be reproduced across dozens of regiments. This mass-produced familiarity, which engendered excitement and loyalty among common soldiers, stands in stark contrast to the experiences of those who closely interacted with Frederick in the course of the war. The future military theorist Georg Heinrich von Berenhorst, at this point a young officer with the king's entourage, recalled:

> Frederick no longer commanded love, respect, or even fear among the nearest and most intimate members of his suite. I can say this because I saw it with my own eyes. When we rode behind him there was a mischievous young brigade-major of the cavalry, called Wodtke, who set out to amuse us by going into comic contortions behind his back, imitating the way he sat in the saddle, pointing at him and so on. Wodtke bestowed on Frederick the nickname 'Grave-Digger'. Later on he abbreviated it to 'Digger', and this is what he called the great hero when we came together in private for jokes and malicious talk.[60]

It is impossible to know exactly how far these sentiments permeated the officer corps, but evidence suggests that they were widespread by late 1761. The British ambassador, Mitchell, commented to Frederick that his officers 'were like an army of Jacobites.'[61] By contrast, this mass-produced familiarity with enlisted men provided one way for Frederick to advance and promote his martial reputation with his soldiers.

Frederick appears to have been widely successful in cultivating sentiments of loyalty among native Prussian troops, and partially successful among non-Prussian foreigners in the army. The first way of measuring this is by examining loyalty in adversity among Prussian prisoners of war. By 1761, Austrian manpower shortages forced Maria Theresa to enlist Prussian prisoners of war against their will. These soldiers deserted the Austrian army in large numbers, and many returned to Prussian service. A non-Prussian, Johann Christian Schimmel, is a perfect example of this type of non-Prussian loyalist. He gained the loyalty of his captors by pretending to convert to Catholicism and preparing to marry a local girl. Upon being given a local suit of clothes, he immediately fled in the night, and over the next few weeks walked some 590 kilometres to his home village of Herzberg in Saxony. Upon returning to Herzberg, Schimmel briefly stayed with his relatives, and then went with his father to meet a Prussian recruiting party, where he was returned to his regiment.[62]

---

60 Berenhorst and Bülow, *Aus dem Nachlasse von Georg Heinrich von Berenhorst*, p.181. Wodkte would eventually travel to the United States and die in the American War for Independence.
61 Johann Georg Zimmermann, *Ueber Friedrich Den Grossen* (Vienna: Ofen, 1788), p.186.
62 Johann Christian Schimmel, 'Kurze Lebensbeschreibung des preussischen Veterans Johann Christian Schimmel,' *Zeitschrift für Kunst, Wissenschaft, und Geschichte des Krieges*, vol.10, no.4–6 (1827), pp.190-192.

Schimmel's Prussian patriotism made him a liability to the Austrians, one that they recognized and tried to mitigate.

Austrian government officials wrote that it would be best to separate Prussian prisoners of war from other deserters and prisoners, 'whom they lead astray by their arguments, and the high opinion they impart concerning their king.'[63] In addition to native-born Prussians, German Protestant foreigners in the Prussian army also retained a deep sense of loyalty to Frederick. A Brunswicker, Johann Heinrich Bittner, spent at least 18 months planning his escape from the Austrian army. Bittner's scheme failed and he was sentenced to 10 years' labour on Austrian fortifications.[64] Austrian army officers also noticed the bonds of loyalty between Frederick and his common soldiers. Jacob Cognazzio, a former Austrian army officer, recalled that Prussian common soldiers knew that there was 'no danger or burden of the war which was not shared by "the Great Fritz" as [Prussian soldiers] called him in childlike respect and love.'[65] Soldiers combined their paternal affection for their monarch with their religious devotion.

Conclusion

This chapter examined negotiated authority common to all eighteenth-century armies, as well as the religious and monarchical ideologies that motivated Prussian common soldiers during the Seven Years War. It argues that Prussian soldiers relied heavily on religious beliefs when attempting to make sense of their lived experiences during this war. As their letters demonstrate, Prussian soldiers believed that a loving and powerful God protected them and their families during this time of crisis. Second, the chapter argues that Frederick II formed an important part of his soldiers' worldview, and he deliberately cultivated his image for his soldiers. The king used Prussian soldiers' loyalty to family and village connections, crafted a mass-produced familiarity with his men, and benefited from a 'God and King' ideology which swept the army in the Seven Years War. Though Frederick was unable to hide his foibles from those who knew him intimately, he was able to be the familial, but stern, *Landesvater* his men expected. These men, rather than mocking and despising Frederick, retained their love and respect for him during the hard years of war. Frederick cultivated a rapport with his soldiers that presented him as a comforting and familiar, yet distant and powerful figure. This combination allowed his soldiers to retain their high opinions of the king and prevented mutiny and scrutiny from the lower ranks during the Seven Years War.

Despite the rancour which existed between Frederick and some of his officers, the 'Great' king was able to retain the loyalty of his common soldiers, whether native Prussian cantonists or foreign volunteers and conscripts. Via a mix of religious

---

63   Kriegsarchiv Vienna: CabinetsAkten 1757 XI 2. Reflections on a Resolution, possibly resulting from the large number of Prussian Prisoners, 26 November 1757
64   KV: Kriegsakten 300, *Sententia* of court martial, 11 February 1761.
65   Jakob de Cogniazzo, *Geständnisse Eines Oesterreichischen Veterans* (Breslau: Löwe, 1791), vol.2, p.27.

solidarity and an established ritual of military paternalism, Frederick personally motivated many of his men, but remained distant enough from them to avoid the familiarity and disdain which marred relations with the Prussian officer corps. As Adam Storring has suggested, other Prussian officers, rather than the genius of Frederick, may be responsible for the victories that the army enjoyed in the Seven Years War.[66] The officer corps seemed well aware of that fact. As Katrin and Sascha Möbius have demonstrated, common soldiers could also criticize the military decisions of the king, particularly when they were called on to attack difficult enemy positions.[67] This forms an important part of negotiated authority on the battlefield: the men were happy to follow the king but believed that he had a duty to try and avoid getting them killed if possible. Despite this, the common soldiers of the army continued to invoke the motto of 'God and the King,' believing these two sources of authority and protection would see them through any dangers. These factors, rather than the external drill so hastily copied by other militaries after the Seven Years War, helped account for the extraordinary successes of the Prussians between 1740 and 1763.

With the concept of negotiated authority and the origins of Prussian success charted, the book will now turn to examining how infantry soldiers from the states of military Europe actually fought on the battlefield between 1733 and 1783. Although Prussomaniac officers argued for rigid control, soldiers took battle into their own hands whenever possible. Soldiers fought in flexible ways that surprised and frustrated military theorists, on the battlefields of the War of Austrian Succession, the Seven Years War, and the American War of Independence.

---

66 Adam Storring, 'Frederick the Great and the Meanings of War, 1730-1755'.
67 Katrin and Sascha Möbius, *Prussian Army Soldiers and the Seven Years War: The Psychology of Honour*, p.147.

6

Obeying the Officers? Aimed Fire and Skirmishing

On 5 July 1759, on the shores of Lake Ontario, British soldiers fought a desperate battle to defend a camp which would soon be built up into Fort Ontario at Oswego, New York. The French forces had tried to surprise the British camp, but the British commander, Colonel Frederick Haldimand, ordered his men to dig in behind barrels and carefully manage their fire. When the dust settled on 6 July, the French withdrew, and the British forces had managed to hold onto their position through a judicious use of aimed fire. A British officer's report reproduced in the *Pennsylvania Gazette* crowed: 'Colonel Haldimand during all the attacks was very active, the enemy, finding all their schemes abortive… could never be prevailed on to run upon us, took to their battoes… it must be told, to the honour of our men, that they behaved uncommonly well, and never fired but when they saw an enemy, and took a very good aim.'[1]

This account, almost certainly written by an officer of the 60th (Royal American) Regiment, provides a unique example of an officer praising his men for what they so often desired. This officer noted that his troops carefully regulated their fire and aimed at the enemy with precision. Taken together, these two attributes meant that the men had, in his words, 'behaved uncommonly well.' Modern authors such as David Blackmore and Alexander V. Campbell have correctly asserted that British troops maintained a preference for delivering fire at close range, with accuracy borne of target practice.[2] This source, providing a clear example of such tactics, raises another point: If the men had truly 'behaved uncommonly well,' this sort of accurate firepower might have been a challenge to maintain, even for the British. This chapter will explore two realms where, to some extent at least, eighteenth-century officers found success in bending the men to their will: utilizing aimed fire and skirmishing tactics.

---

1   'Camp at Lake Ontario, July 7, 1759', *The Pennsylvania Gazette,* 2 August 1759.
2   Blackmore, *Destructive and Formidable,* passim; Campbell, *The Royal American Regiment: An Atlantic Microcosm, 1755–1772*, pp.80–119.

Aimed Fire: Its Proponents and Challenges

One of the major debates over the use of fire in the eighteenth century concerned whether to emphasize firing at speed or firing slowly and accurately. A British manual for militia after the Seven Years War chided its readers:

> The principal article is, for the officers to see that the men take a good aim… to keep up a steady fire, well aimed and directed, so as to do execution, than fire away double or treble the number of shot, in the same time, to little or no purpose: which is an extreme we seem to have ran into of late years; though so late authors and authorities have assured us, that it is not the very quick fire of the Prussians that has the most contributed to their victories…[3]

Despite the debates over speed of fire and accuracy, most European armies desired accurate fire from their men, and officers attempted to facilitate this accuracy. In 1755, James Wolfe described the process of training his men to a fellow officer. His letter took the form of giving advice on the advantages of target practice:

> We fire bullets continually… and let me recommend the practice, you'll soon find the advantage of it. Marksmen are nowhere so necessary as mountainous country, besides, firing balls at objects teaches the soldiers to level incomparably, makes the recruits steady, and removes the foolish apprehension that seizes young soldiers when they first load their arms with bullets. We fire… and the soldiers see the effects of their shot especially at a mark, or upon water. We shoot obliquely, and in different situations of ground from heights downwards and contrarywise.[4]

Officers were insistent in their recommendation for aimed fire, and this practice was adopted by other armies fighting in military Europe. *Subsidientruppen* allied with the British in the American War of Independence utilized this practice as well. The leader of the Braunschweiger forces in Canada, Riedesel, noted:

> General Carleton has decreed that that the army is to practice target shooting or shooting at a [mark] I am issuing orders so as to adhere to this as well and as successfully as possible. Each squadron or company is to have some rough boards fastened together, on which a ring of black is to be painted proportionately. The target may be either square or round. Have this target placed at a distance of point-blank shot.[5]

---

3   William Windham and George Townshend, *A Plan of Discipline for the use of the Norfolk Militia* (London: Millan, 1768), p.190.
4   Wilson, *The Life and Letters of James Wolfe*, p.255.
5   'Brunswick Order Book, 1776–83,' in *Hessian Documents of the American Revolution* (Boston: G.K. Hall, 1989), p.H.Z. 21, microform.

This might seem to be a contradictory point, as today the term 'point blank' has taken on the colloquial meaning of extremely close. Fortunately, we have a good idea of what this term meant in an eighteenth-century context. Lewis Lochee, a military intellectual from the Austrian Netherlands who ran a military academy outside London, noted in 1783, 'the point-blank of our firelocks… is known to be about 300 yards.'[6] This is quite a long range for target practice, it is possible that Riedesel meant 250 yards or closer. At the very least, then, Riedesel intended for his men to practice at the regular distance of an eighteenth-century firefight.[7] With the distance of the target practice established, Riedesel continued:

> Each man of the company will shoot eight times… a non-commissioned officer standing some distance from the target will mark on the target the location of the shot… the company officers are to gather together to prove to the men how well they can shoot and aim. When each man has shot eight times, the entire target practice will be completed. In the judgement of the captain, the one who has made the best four shots will receive the following prizes…[8]

Riedesel goes on to list a monetary reward, and also gives instructions that men who are waiting their turn to shoot should be kept at ease, so as not to think of target shooting as a hard duty, but something necessary for military life. If German *Subsidientruppen* in North America took to target shooting with such alacrity, we should not be surprised to find that German soldiers in Europe also worried about the accuracy of their shots.

The Prussian infantry regulations of 1750 instructed soldiers to '…take aim along the barrel with the sight, and look boldly into their fire. The soldier must know where he is shooting, most importantly, not too high or too low, and the officers must carefully watch for this.'[9] The regulations assured that the best way to take a good aim was to 'keep the [musket] but level with the shoulder, and the men should sink their heads down, so they can see where they are shooting.'[10] The regulations then expounded: 'The officers must ensure that the men have a good stance before they order them to present, that they level well, and that they take a good aim.'[11] A note later in the regulation giving advice to officers who were attacked while isolated from the main army indicated they should 'fire always regularly by platoons, and not while the enemy is at too great a distance, and they take a good aim.'[12]

---

6 Lochee gives some indication that it might be shorter than this distance but references 250 yards and 300 yards subsequently in the treatise. Lewis Lochee, *Elements of Field Fortification* (London: Cadell, 1783) pp.68, 87, 135.
7 See chapter six.
8 'Brunswick Order Book, 1776–83', p.H.Z. 21, microform.
9 *Reglement vor die Königlich Preussische Infanterie* (Berlin: Given and Printed 1 June 1750), p.57.
10 *Reglement vor die Königlich Preussische Infanterie*, p.57.
11 *Reglement vor die Königlich Preussische Infanterie*, p.61.
12 *Reglement vor die Königlich Preussische Infanterie*, p.256.

During the 1740s and early 1750s, it seems as though the Prussians stressed aimed fire more than the British, who became unique in their emphasis on aimed fire as the century wore on. Thomas Simes argued in 1777 that soldiers should be forced to practice firing 'with ball at the target till they can hit the object six times out of twelve.'[13] Simes then quoted Major Thomas Bell's 1770 essay on first military principles, which in full reads: 'Without firing at a mark, men will not be marksmen; and without being sure to kill, soldiers are not in the best possible state for war. A battalion whose fire is certain and deadly kills, stops, and conquers; a battalion whose fire is unsure, is unkilling, will not stop, and many be conquered.'[14]

In other portions of his essay, Bell ranted, 'could any person imagine that European soldiers were armed with firelocks, yet that those men were not trained to be marksmen, or that they never, or very rarely, fired at a mark?'[15] French officers similarly called for soldiers to be taught to aim, as the comte de Guibert noted in 1772.[16] Thus, George Washington joined a chorus of commanders who called for aimed fire. He noted in 1782:

> The officers commanding the light infantry should impress upon the men the necessity of taking deliberate aim whenever they fire and see that they do it when it is in their power – It is the effect of the shot not the report of the gun that can discomfort the enemy and if a bad habit is acquired at exercise it will prevail in real action and so vice versa.[17]

American soldier Joseph Plumb Martin gave the clearest description of employing aimed fire in combat. While acting as a skirmisher for the Continental Army at the Battle of Monmouth in 1778, Martin recalled:

> When within about five rods of the rear of the retreating foe, I could distinguish everything about them, they were retreating in line, though in some disorder; I singled out a man and took my aim directly between his shoulders, (they were divested of their packs) he was a good mark, being a broad shouldered fellow; what became of him I know not, the fire and smoke hid him from my sight; one thing I know, that is, I took as deliberate aim at him as ever I did at any game in my life. But after all, I hope I did not kill him, although I intended to at the time.[18]

---

13 Thomas Simes, *A Military Course for the Government and Conduct of a Battalion* (London: Almon and Hooper, 1777), p.177.
14 Thomas Bell, *A Short Essay on Military First Principles* (London: Becket and Hondt, 1770), p.2.
15 Bell, *A Short Essay on Military First Principles*, pp.116–117.
16 Guibert and Abel (trans.), *Guibert's General Essay on Tactics*, pp.74–75.
17 LOC: Washington Papers, General Orders, 8 June 1782.
18 Joseph Plumb Martin, *A Narrative of some Adventures, Dangers, and Sufferings of a Revolutionary Soldier* (Hallowell: Glazier, Masters & Co, 1830), p.94.

Roger Lamb recalled seeing American troops at Guilford Courthouse 'taking aim with the nicest precision.'[19] It was not only American troops who aimed: William Dansey of the 33rd Regiment noted, 'A fellow… ran behind a tree and presented at me; I up with my Fuzee [officer's musket] knocked him as quick as a cockrooster would a cock.'[20] Although Dansey was quite forthright about the act of shooting another human being, most eighteenth-century soldiers had more qualms about shooting at a specific target.

John McCasland, a rifle-armed Pennsylvanian in the American War of Independence, gave the fullest expression of this sentiment when asked to shoot down a Hessian sentry in 1778:

> We cast lots, and it fell to my lot to shoot the Hessian. I did not like to shoot a man down in cold blood. The company present knew I was a good marksman, and I concluded to break his thigh. I shot with a rifle and aimed at his hip. He had a large iron tobacco box in his breeches pocket, and I hit the box, the ball glanced, and it entered the thigh and scaled the bone of the high on the outside.[21]

Most historians who have examined this passage emphasized McCasland's unwillingness to shoot a sentry, but his exact phrase was: 'I did not like to shoot a man down in cold blood.'[22] This matches well with the self-forgiveness that Marian Füssel has identified this type of sentiment, an absolution of personal responsibility for harming others in combat, as typical in the writings of eighteenth-century soldiers.[23] Prussian soldier Ulrich Bräker described the strain of combat in 1756 at Lobositz in a similar way, stating: 'I…fired away all my sixty cartridges, so that my musket became so hot I had to carry it by the sling. But I don't think that I hurt a living soul, it all went into thin air.'[24] This expression also parallels the reminiscences of the American soldier Martin: 'But after all, I hope I did not kill him, although I intended to at the time.'[25] Even if soldiers had desired to take a cool and deliberate aim at every shot, other factors impeded their progress.

---

19 Lamb, *An Original and Authentic Journal of Occurrences during the Late American War*, p.361.
20 Historical Society of Delaware: Letters of Captain William Dansey, William Dansey to his mother, 15 March 1777.
21 John C. Dann, *The Revolution Remembered: Eyewitness Accounts of the War for Independence* (Chicago: University of Chicago Press, 1980), p.156.
22 Christopher Duffy, *Military Experience in the Age of Reason* (London: Routledge & Kegan Paul, 1980), p.163; Jon Chandler, 'The Continental Army and "Military Europe": Professionalism and Restraint in the American War of Independence', *War in History*, vol.29, no.2 (2020), p.331.
23 Marian Füssel, 'Emotions in the Making: The Transformation of Battlefield Experiences during the Seven Years War' in Erika Kuijpers and Cornelius van der Haven (eds), *Battlefield Emotions 1500-1800: Practices, Experience, and Imagination* (London: Palgrave Macmillan, 2016), p.153.
24 Ulrich Bräker and Johann Heinrich Füssli, *Lebensgeschichte und natürliche Ebentheuer des Armen Mannes im Tockenburg* (Zürich, Orell, Gessner, Füssli,1789), p.150.
25 Martin, *A Narrative of some Adventures, Dangers, and Sufferings of a Revolutionary Soldier* (Hallowell: Glazier, Masters & Co, 1830), p.94.

The discharge of thousands of weapons filled eighteenth-century battlefields with smoke. In an age before smokeless powder, the battlefield rapidly became hazy. The cumulative effect of the mass firing of a battalion was to quickly hide their target from view. If the enemy was also firing back, this only compounded the situation. The British soldier John Robert Shaw described the scene the Battle of Camden in 1780: 'The haziness of the morning prevented the ascent of the smoke, which occasioned such a thick cloud that it was difficult to observe the effects of a well-supported fire on both sides.'[26] George Washington noted the same at Germantown in 1777.[27] Charles Immanuel de Warney noted that one of the principal challenges to effective firepower was 'the blinding and suffocating smoke.'[28] Some soldiers reported that despite the smoke, they were still able to 'generally see the enemy,' even if there was a 'great deal of smoke,' while implying that the enemy seemed to fade in and out of view.[29] A French officer remonstrated that even if muskets were better designed so that soldiers could aim, 'in heavy action… a thick smoke hides their target.'[30] Smoke combined with tactical terrain to obscure bodies of men on the battlefield, as a Prussian soldier found at Zorndorf in 1758:

> I can only say what I saw myself. The dust was quite strong. We came to a large ditch: some crossed through it, some jumped it, others found a narrow path around it. Beyond this trench we found twenty Russian Cuirassier who were defeated by our cavalry. Then we stumbled upon a regiment of Prussian infantry who were marching back with reversed arms… we saw wounded men on the battlefield… I saw the unfortunate wounded with great sadness. By that point, we had advanced half a mile, and stopped in the village of Bikker. We searched about, and when the dust had finally cleared, half of our unit was missing.[31]

In 1760, a Prussian regiment came under enemy fire during a skirmish but reported, '…since we were a distance away and marching quickly, they didn't do us much harm, which was due to the fact that there was a heavy haze, which prevented them from seeing us properly, and they usually over or undershot us.'[32] Even if enemy troops were roughly visible, clearly aiming through the haze of dust, smoke, or even morning fog could present a challenge to marksmen.

---

26 John Robert Shaw, Oressa M. Teagarden, and Jeanne L. Crabtree, *John Robert Shaw: An Autobiography of Thirty Years* (Athens: Ohio University Press, 1992), p.31.
27 LOC: Washington Papers, General Orders, 5 October 1777.
28 Charles Immanuel de Warney, *Remarques sur plusieurs auteurs militaires et autres* (Lublin: Staroludzki, 1782), pp.69–70.
29 Reverend Augustine G. Hibbard, *History of the Town of Goshen, Connecticut* (Hartford: Case, Lockwood & Brainard, 1897), p.145.
30 Anon., *La Foiblesse du feu trop precipité, du canon et du mousquet* (Liege: Jaques Balbin, 1759), p.8.
31 Friedrich Adolph von Kalckreuth, 'Erinnerungen des General-Feldmarschalls Grafen von Kalkreuth,' *Minerva: Ein Journal historischen und politischen Inhalts*, vol.4 (1840), p.146.
32 Anonymous, *Sammlung ungedruckter Nachrichten so die Geschichte der Feldzüge der Preußen von 1740 bis 1779*, (Dresden: Walther, 1785) Volume 2 p.431.

How accurate was the aim of eighteenth-century soldiers armed with smooth-bore weapons? Although lacking rifles and firing weapons that produced vision-obscuring smoke, eighteenth-century soldiers were repeatedly instructed to fire accurately across military Europe. The idea that Revolutionary American soldiers were the first to aim their weapons due to the phrase 'take aim' being included in Friedrich Wilhelm de Steuben's drill book is an essentialist myth. Troops from all over central and western Europe engaged in target practice and were instructed to carefully aim their weapons by their officers throughout the first half of the eighteenth century. If the Americans were exceptional to a degree, it was thanks to a pre-existing British emphasis for aimed fire that included more extensive live-ammunition target practice than most contemporary militaries. If these soldiers certainly tried to aim their weapons, they also fought in ways that might surprise those who view the eighteenth century as a benighted age of strict linear warfare.

Origins of the Flanquer System

When thinking about skirmishers, those alive to a sense of the past often locate their development in the Napoleonic era, with the French *tirailleurs,* and their quick emulation by all of the European states. In fact, it appears that a form of skirmisher developed during the eighteenth century. These troops were often called 'flanquers,' 'flanqueur,' sometimes 'Blänkerer' or at times, even *tirailleur,* by French and German speakers, or Flankers, by the English speakers. This can be confusing, as the same term was often applied to men who were assigned to security duty to the sides of columns during a march. This idea of skirmishing is often controversial, as many military history enthusiasts believe that eighteenth-century soldiers were not trustworthy enough to fight in this fashion and that they would desert if fighting in this style. It is important to note: the use of the term 'skirmishers' in this chapter is not referring to dedicated light infantry or ranger units, but the use of integrated skirmishers by regiments of infantry and cavalry. While these soldiers were not universally employed, it is possible to locate them before the French Revolution.

The idea that the Hessians used skirmishers, (not Jäger, but integrated line skirmishers) supports connections between the Hessen-Kassel and Prussian military systems. Atwood indicates that the Hessians copied their drill manual from the Prussians, which when it comes to speed of fire may well be right.[33] By the 1780s, Frederick the Great argued that *Frei-Infanterie* should be adept at taking cover 'behind trees… in houses…lie flat…shoot from behind stones…fire from behind the crest of a ridge.'[34] Frederick may or may not have intended to use *Frei-Infanterie* as skirmishers, depending on how one reads the 1783 *Instruction*.[35] However, another

---

33 Atwood, *Hessians,* p.66.
34 Gustav Berthold Volz, *Die Werke Friedrichs des Grossen in deutscher Übersetzung*m (Berlin: Reimar Hobbing, 1913), p.300.
35 Duffy, *The Army of Frederick the Great* (1996), p.138.

Prussian commander may well have used infantry skirmishers during the Seven Years War.[36]

A Hungarian officer reported that at the Battle of Mollwitz, 'Our infantry had advanced a platoon of men in front of each battalion, to attack the enemy first, and be supported by us, yet this was not observed, and robbed us of the only means of striking such as well-trained enemy.'[37] There are other examples of such detached platoons, but it is not clear if these men were operating in loose order.[38]

At the Battle of Reichenbach on August 16th, 1762, the Duke of Bevern employed infantry *Pelotons* (platoons) as skirmishers in order to disturb the Austrian advance. Two writers from the period noted this use of individual Prussian *Pelotons* as skirmishers, in order to delay the advance of the Austrian column under *Generalleutnant* Beck.

The first author to notice this strange use of the normally rigid Prussian infantry was Georg Friedrich Tempelhoff, who had served as a Prussian gunner during the war and translated and expanded the work of Henry Lloyd, a veteran of the Austrian forces.[39] Tempelhoff describes the Duke of Bevern's use of these *Pelotons* at Reichenbach in the following way:

> Because the Duke was going to have a second engagement, and could not draw from the forces already engaged without showing the enemy his weakness, he drew off a few platoons (*Pelotons*) and put them into the swampy area by the Schrober-Grund. These platoons fired on the enemy, and resisted during his march, in order to delay him as long as possible.[40]

Tempelhoff also noted a higher number of causalities suffered by these platoons, indicating that they were in heavy contact with the enemy.

The description of a single source, even one as respected as Tempelhoff, might not be enough to confirm the use of *Pelotons* as a delaying force at Reichenbach. However, another military authority, Hessian Jäger Johann von Ewald, in the third volume of his work, *Belehrungen über den Krieg: Besonders über den kleinen Krieg*, described the same type of occurrence at the Battle of Reichenbach. Ewald knew officers who had served in the Prussian army and was himself a veteran of the Seven Years War in Western Germany. He described the Duke of Bevern's reaction to the appearance of Beck's corps, saying:

> In this battle, the Prussian general was obliged (as a strong as a strong Austrian corps under General Beck bypassed the Prussian left, and had already manoeuvred behind and above the Prussians, while the Prussian

---

36 Atwood, *Hessians*, p.83.
37 Anon., *Sammlung ungedruckter Nachrichten*, vol.1, p.35.
38 Anon., *Sammlung ungedruckter Nachrichten*, vol.5, p.568.
39 Christopher Duffy, *By Force of Arms: The Austrian Army in the Seven Years War Volume 2* (Chicago: Emperor Press, 2008), p.473.
40 George Friedrich Tempelhoff and Henry Lloyd, *Geschichte Des Siebenjährigen Krieges in Deutschland*, (Berlin, 1801), vol.6, pp.149–150.

front was already engaged with the enemy) to draw off single platoons of different battalions, and send them backwards to offer resistance to the enemies' lead elements. As a result of this, and also the appearance of the army of the king, the enemy was completely repulsed, which decided the fate of the battle and the whole campaign.[41]

From these two examples, it is clear that not only did the Prussians definitely use individual platoons in this battle, but that individual platoons were often used to screen larger infantry battalions. A soldier of the Reichsarmee recalled that in 1762, 'Our flanqueurs engaged the enemy without orders.'[42]

By the early 1770s, it had become commonplace to refer to groups of infantry fighting ahead of the main body as 'flanquers.' Reporting on ongoing Russian war against the Ottoman Empire, a newspaper in 1773 described a Russian manoeuvre, saying, 'in order to protect the encampment, volunteers from the infantry were added to the flanquers.' Continuing the story of a battle against the Ottomans, the newspaper concluded, 'the right-hand square was attacked with such strong force that both the light troops and the flankers on foot (*Flanquers zu Fuß*) were driven back.'[43]

The Prussians may have employed skirmishers, as Frederick II of Prussia gave detailed instructions for an 'officer who is… to cover an army or regiment whilst they are deploying... must send out flankers towards the enemy, who, by keeping up a constant firing, will endeavour to disperse them'.[44] Based on the context, it is very likely that this instruction is meant specifically for cavalry officers. Prussian cavalry flanquers were noted both by contemporary authors and subsequent historians.[45] It seems at that times, cavalry flanquers were used to screen attacking infantry.[46]

It is unclear whether Prussian line-infantry regiments utilized flanquers in the Seven Years War. Rather, it seems that the Prussian infantry may have used *Heckenfeuer,* what Christopher Duffy has called a type of 'controlled skirmishing.' In this process, two files advanced ahead of the regimental body, formed in two ranks, fired, and then retired to the main body while reloading. This process enabled the battalion to keep up a small but consistent rate of fire, while retaining a reserve of loaded muskets.[47]

---

41 Johann von Ewald, *Belehrungen übber den Krieg: Besonders über den kleinen Krieg, durch beispiele großer Helden und Kluger und tapferer Männer* (Schleswig: Röhß, 1803), vol.3 p.321.
42 Anon., *Beyträge zur neuern Staats-Und Krieges-Geschichte* (Danzig: Schuster, 1762), vol.16, p.335.
43 'Mittewoch, vom 25sten August 1773,' *Beytrag zum Reichs-Postreuter*, Issue 136, p.4.
44 Frederick II and T. Forster, *Military Instruction from the Late King of Prussia to His Generals* (Sherborn: J. Cruttwell, 1818), p.114.
45 Friedrich Wilhelm von Gaudi and Martin Löffelholz, *Journal vom Siebenjährigen Kriege* (Hamburg: LTR Verlag, 1996), vol.7, p.220; Achim Kloppert 'Der Schlessische Feldzug von 1762' (unpublished doctoral dissertation Universität Bonn, 1981), p.404.
46 Henri de Catt and Reinhold Koser, *Unterhaltungen mit Friedrich dem Großen: Memoiren und Tagebücher* (Leipzig: Hirzel, 1884), p.359.
47 Duffy, *The Army of Frederick the Great*, (2020), pp.144–145.

On 3 March 1777 or 1778 (the source is unclear), the Prussian cabinet ordered 10 men per infantry company (an administrative rather than tactical formation) to be drilled as skirmishers in order to serve in patrols and detachments.[48] Charles Immanuel de Warnery, a Prussian Hussar officer, noted that one of the principal duties of the Prussian Jägers was to 'keep enemy flankers at a distance.'[49] In September of 1778, Prussian observers reported encountering enemy 'flanqueurs' on their way to a hilltop position.[50] In his *Geschichte des Kriegskunst*, published in 1800, Johann Gottfried von Hoyer asserted that the German 'flanqueur,' like the Volontair and the Jäger, had preceded the *Tirailleur* of the French Revolution.[51]

However, the Hessians did not serve in the eastern German theatre where the Battle of Reichenbach occurred – they were busy fighting the French in the western Holy Roman Empire. Thus, while military thought in Hessen-Kassel was influenced to a great extent by the successes of the Prussian army, it remained a distinct entity. In order to fully understand the development of the skirmish system employed by the Hessians in the American War of Independence, it is necessary to understand the use of light troops in the Seven Years War and specifically how these troops were employed by the French.

Many authors, such as Reginald Savory, have described the rise of light troops within European armies in the Seven Years War. However, it is generally asserted that these troops played a minor role in set-piece battlefield encounters. There is evidence to suggest that the French may have begun using integrated line skirmishers, particularly towards the end of the Seven Years War in small encounters. At the attempt on Lippstadt, on 1 July 1759, the comte de Melfort described his use of skirmishers in a small confrontation with Hanoverian troops.

In the battle, the French used skirmishers in an attempt to clear the way for an attack. In his letter, Melfort uses the term *Tirailluers* – the traditional French term for skirmishers which we commonly associate with the French Revolutionary Wars. A French dictionary from 1752 defines a 'Tirailleur' as 'one who skirmishes.'[52] Melfort's choice of this term, and the forces present at the battle, indicate that he is referring to line infantry skirmishers, not *Chasseurs* or *Volontaires*. Thus, according to the letter of the comte de Melfort, both the French and the Hanoverians used skirmishers in this conflict.[53]

The idea that the French employed skirmishers in the Seven Years War is confirmed by a letter from Victor-Francois, duc de Broglie. In this letter, Broglie confirms that in the winter of 1759–1760 French infantry regiments trained 50 men per battalion to operate as skirmishers.[54] The 1764 *Ordonnance du Roi*, which was

---

48   Pascal Bressonet, *Études tactiques sur la campagne de 1806* (Paris: Imprimerir R. Chapelot, 1909), p.371.
49   Charles Emmanuel de Warnery, *Remarks on Cavalry by the Prussian Major General Warnery* (London: Barfield, 1798), p.106.
50   Friedrich Wilhelm von Schmettau, *Mémoires raisonnés sur la campagne de 1778 en Bohème* (Berlin: 1789) pp.186–187.
51   Johann Gottfried von Hoyer, *Geschichte der Kriegskunst* (Göttingen: Rosenbusch, 1799). p.941.
52   Annibale Antonini, *Dictionnaire françois, latin et italien* (Venice: François Pitteri, 1752), p.520.
53   Services Historique de l'Armée de Terre: A1 3518, pièce 40, p.3.
54   Jean Lambert Colin, *L'Infanterie Au XVIIIe Siècle. La Tactique* (Paris: Berger-Leverault, 1907), pp.76–77.

likely written by de Broglie, the French Infantry are instructed to use skirmishers: 'Nothing should prevent you, when on the advance or retreat, from detaching a half-section, and scattering these volunteers in front of the battalion, to make a feu de billebaude, and then retreat through the intervals behind the battalion when the enemy is very close.'[55]

The British also seem to have followed this practice in an impromptu way. In advance of the main British line at Vellinghausen in 1761, Corporal William Todd reported, 'the enemy had gotten a large body of troops...we kept a heavy fire upon them. Two very nimble sergeants belonging to the highlanders skipped from tree to tree near the enemy and fired several shots.'[56] In both practice and theory, we have men from regular infantry units, rather than light troops, being utilized in front of the main body of a battalion of infantry in order to delay the enemy and weaken them before they are confronted by the main body of the battalion.

The fullest description of skirmishers by an old regime officer was provided by the French writer Knock in 1779. He recommended detaching a platoon of skirmishers from the battalion in order to cover the front of the battalion, who could withdraw to the battalion itself when the enemy came close enough to truly be damaged by the fire of the whole battalion. He then answered several possible objections to skirmishers, including how they might mask the fire of friendly artillery, and that while the skirmishers were deployed, the men of the battalion essentially had nothing to do and could not fire on the enemy. Here, Knock argued, was the greatest advantage of all: the skirmishers would essentially keep the battalion from firing without orders. He argued, '...discipline would have to be extremely lax, or the troops would have to be the worst sort of cowards by firing before the retreat of the skirmishers.'[57]

The French and western Germans both developed a plethora of light troops for use in the war of posts. In addition to this development of light troops, the use of integrated skirmishers appears to have been in the extremely early stages of development, and this development was noticed and emulated by the western German army facing the French, under Ferdinand of Brunswick. In a letter describing the Battle of Bergen, fought on 13 April 1759, Ferdinand of Brunswick described the early stages of the action: 'I ordered our Grenadiers and Jägers to amuse the enemy, supported by detached platoons, so that our columns would have time to arrive.'[58] The three types of troops used in this delaying action correspond exactly to the three types of troops used that the Hessians would use in the Flatbush Pass on Long Island in 1776. When all of this data is collected, it appears the skirmishing system employed by the Hessian *Subsidientruppen* developed out of the Seven Years War in Europe, rather than their North American experiences.

In the American War of Independence, the Hessian forces repeatedly used skirmishers. In his diary, *Premierleutnant* Jakob Piel noted that at the onset of the

---

55 *Ordonnance du roi, pour régler l'exercice de l'infanterie. Du 20 mars 1764* (Paris: De L'Imperimerie Royal, 1764) pp.106–107.
56 Todd, *The Journal of Corporal William Todd*, p.164.
57 G. Knock, *Lettres militaires* (Paris: Duchene, 1779), p.91.
58 Ferdinand Von. Westphalen, *Geschichte Der Feldzüge des Herzogs Ferdinand Von Braunschweig-Lüneburg*, (Berlin: Verlag der König, 1872), vol.3, p.242.

attack at Flatbush, 'Lieutenant Zoll with fifty volunteers was sent into the woods and exchanged shots with the rebels.'[59] Piel does not seem to think that this practice of selecting volunteers was anything out of the ordinary. Piel uses the term volunteers to describe these men, indicating that they were not part of a pre-formed group. The official journal of the Alt-Lossberg regiment notes the event in similar terms.

Another young officer, Karl Friedrich Rüffer of the Mirbach Regiment, described his role in the action: 'Lieutenant Schraitd was sent into the thicket with some eighty flanqers, which resulted in us taking some captives.'[60] Thus, Rüffer described the activity of his men by using the term of *Flanquer*. Again, these are men drawn from the main body of the regiment, not dedicated light infantry.

Johann Heinrich von Bardeleben of the von Donop regiment was also present and described the attack: 'Because of the landscape and the terrible hills, [the rebels] could not be attacked *en masse*, but only by groups. All the regiments at once sent out strong patrols, as strong as possible, in general, to attack the enemy.'[61] Bardeleben's description presents two problems for the interpretation that this attack was a regular practice. First, he indicated that the only reason the Hessians sent out skirmishers was 'because of the landscape,' and second, he referred to them as 'patrols,' not *Flanquers* or skirmishers. Let us address these concerns in detail.

First, while the Hessians were undoubtedly motivated by the rough nature of the terrain they had to attack, they would continue to use this formation, even when the position they were attacking was not in rough terrain. Examples of this will be seen later in the chapter specifically at the Battle of Trenton. Next, while Bardeleben specifically referred to them as patrols, most other Hessian officers, including Jäger veteran von Ewald and the Hessian commanding general von Heister, referred to this type of soldier as a *Flanquer*.[62] The Hessian use of *Flanquers* was extremely successful at the Battle of Long Island, and the Hessians were able to apply the appropriate amount of pressure, which allowed the British to devastate the Rebel positions.

The overall Hessian commander, von Heister, was less than impressed with the British ideas regarding swift-movement in open order.[63] In a report to Landgraf Friedrich II of Hesse-Kassel, Heister described his thoughts on the skirmishing order which the Hessians used, and its comparison to the British system: 'Because of the steady fire of the Jäger, henceforth, the four Grenadier battalions, brigaded together with the English Grenadiers and Light Infantry, will obey the regulations of General Howe, to which the troops of the army submitted.' Here, Heister agreed that since his combined Grenadier battalions were brigaded with the Jäger, British

---

59   Jakob Piel, Andreas Wiederholdt, and Bruce E. Burgoyne, *Defeat, Disaster and Dedication: The Diaries of the Hessian Officers Jakob Piel and Andreas Wiederhold* (Bowie, MD: Heritage Books, 1997), p.17.
60   August Schmidt, Karl Friedrich. Rüffer, and Bruce E. Burgoyne, *The Hesse-Cassel Mirbach Regiment in the American Revolution* (Bowie, MD: Heritage Books, 1998), p.54.
61   Johann Heinrich von Bardeleben et al., *The Diary of Lieutenant von Bardeleben and other von Donop Regiment Documents* (Bowie, MD: Heritage Books, 1998), p.56.
62   Johann Von Ewald, *Diary of the American War: A Hessian Journal* (New Haven: Yale University Press, 1979), p.85; Hessisches Staatsarchiv Marburg (HSM): 4h. 409 nr.3, fol.53.
63   Atwood, *Hessians*, p.82.

Grenadiers and Light Infantry, it would make sense for them to follow the new British open order system.[64]

However, von Heister refused to be swayed when it came to the infantry under his personal command. He went on to describe the system which the other troops of the Hessian army had been using, and would continue to use:

> However, the rest of the infantry perform honourable service, which is evident from the reports of more than a few regiments. The platoons of skirmishers peel out, at all times gives the best service; but the main battalions are always closed up arm in arm, following the skirmishers at a musket shot distance, unless the rough terrain forces them, at some times, to break ranks, which the reports show is happening rather often.[65]

For von Heister, this system provided the best of both worlds.

On 26 December 1776, the rebel army under the command of George Washington attacked the Hessian garrison outpost at Trenton. In popular memory, the Americans defeated the Hessian garrison by skill and surprise because the Hessians had been drunk after a Christmas celebration. Historian David Hackett Fischer destroyed this myth in his landmark study of the battle, *Washington's Crossing*. In this book, Fischer clearly demonstrates that not a drop of liquor had been drunk before the battle, and that the Hessians were worn out from days of patrols in the snow and rain. Fischer displays that some 6,000 Americans, with a vast superiority in numbers and cannon, easily defeated the 1,400 Hessian defenders.[66]

If the Hessians were sober and alert, did they deploy their integrated skirmishers? While this idea has been missed by most historians who have studied the battle, there is evidence to suggest that the Hessians did indeed attempt to use their skirmishers during the battle. Hessian *Sekondleutnant* Christian Sobbe described the battle. Near the end of the conflict, the von Knyphausen regiment attempted to escape from American capture by fleeing across the river. Sobbe indicates that the von Knyphausen regiment employed skirmishers at this point in the battle. Sobbe states, 'the regiment had been driven into a marsh, [the edge of the river] and stuck fast there, and the regiment had been halted.' Sobbe continues, 'Captain von Biesenrodt had thereupon called out, "skirmishers to the fore."' For Sobbe, this meant a personal instruction. As a lieutenant, this order instructed him to 'run around the battalion in order to call out the skirmishers required, and to form them up.' This allows historians to understand, with certainty, that the command to call out *Flanquers* came from the commander of the regiment, who was in this case *Hauptmann* von Biesenrodt, because *Major* von Dechou had been wounded. Once the command had been issued by the regimental commander, lieutenants were responsible for gathering up the skirmishers.[67]

---

64   HSM: 4h.410 nr.1 fol.507, Heister zu Landgraf, 21 March, 1777.
65   HSM: 4h.410 nr.1 fol.507, Heister zu Landgraf, 21 March, 1777.
66   David Hackett Fischer, *Washington's Crossing* (Oxford: Oxford University Press, 2004), pp.225–235, 392–395.
67   'The Affair at Trenton,' in *Hessian Documents of the American Revolution* (Boston: G.K. Hall, 1989), p.M.L.375, microform.

In addition, it was the responsibility of the lieutenants to lead the skirmishers forward. *Sekondleutnant* Werner von Ferry of the von Knyphausen regiment also described this moment in the Battle of Trenton when the Hessians attempted to use skirmishers. Von Ferry states, 'As the regiment reached the water, Captain von Biesenrodt had ordered, "skirmishers forward."' On that command, von Ferry 'marched forward with the skirmishers up the hill.' Upon reaching the hill, von Ferry notes that they encountered resistance: 'The skirmishers were fired on from the woods by the militia, and immediately... Lord Sterling and his brigade... had marched up on this same hill.' Von Ferry indicates that, 'because the enemy were pressing... in such great strength,' he and the skirmishers 'had been forced to retire to the regiment.' Von Ferry's testimony shows an operation of Hessian skirmishers from beginning to end: the order came to deploy, the men marched forward and encountered resistance, and when that resistance mounted, they returned to the shelter of the regiment.[68]

Not content to let his junior officers tell the story, *Hauptmann* von Biesenrodt also described his role in sending skirmishers forward at Trenton. Biesenrodt stated, 'The skirmishers that I sent forward had begun to fire, but very few of the muskets would go off, owing to the heavy snow and rain.' A little further on, Biesenrodt indicated, 'The enemy had pressed on in such great numbers,' that the skirmishers had been forced to retire. While his description ends there, Biesenrodt was later recalled giving further testimony, when he informed the court of the reasons for his decision to deploy skirmishers.[69]

In this second testimony, Biesenrodt gives insight into the thinking which led him to deploy skirmishers, and what his goal in doing so had been. As the von Knyphausen Regiment was caught between the Americans and a marshy creek, he had been attempting to find a ford, by which the men of the regiment could escape. Thus, in his words, his 'intention had been cover the regiments crossing through the water with the skirmishers, and... to make with them also the rear guard.' Sadly, Biesenrodt noted that this had been made 'impossible by the rapid approach of the overpowering enemy.' From this testimony it is apparent that the deployment of skirmishers could not only be used to cover the regiment during an attack, but also to cover a regiment during withdrawal.[70] By this point, the fact that the *Subsidientruppen* used skirmishers in some, but not all, of their battles in North America has been established beyond doubt. This has already been documented by other historians, if never with this amount of detail.[71] Like their development in Europe, the employment of these skirmishers formed one of the intersections where enlisted men's desire to fight in a more flexible manner connected with their officers' desire to innovate. As a result, the men obeyed their officers.

---

68 'The Affair at Trenton,' p.M.L.444.
69 'The Affair at Trenton,' p.M.L.456.
70 'The Affair at Trenton,' p.M.L.481.
71 Indeed, this style of fighting is not mentioned by Friederike Baer in her excellent new survey of the *Subsidientruppen*. Friederike Baer, *German Soldiers in the American Revolutionary War* (Oxford: Oxford University Press, 2022).

In the American War of Independence, the British used 'skirmishers' in specific instances. Authorities, such as Matthew H. Spring, assert that they did not, and we must grapple with this view.[72] Spring asserts that the light companies formed before the war were not employed as skirmishers, but rather used to form assault battalions which fought in the same, two-rank, open-order style as the rest of the British infantry forces. The fullest vision for the light companies given before the war was George Townsend's orders to the light infantry of the Irish Establishment given in 1771. In this document, a full and complete picture of the duties of light infantrymen is given:

> The Light Infantry must also be taught to take Advantage of large stones , broken Inclosures , old Houses , or any strong feature which presents itself upon the face of a Country . But they must take particular Care not to run in Crowds to these objects … The Arms of every Soldier should be always kept in good Order , But the Light Infantry Man , in particular , must not neglect his Arms , his Ammunition or throw away his Fire , as his Existence may depend upon a Single Shot's taking place . The Light Infantry must consider that the Service upon which they are likely to be Employed , is very different from that of heavy Troops , The former being always to Engage in open Order and the Attack may frequently become personal between Man and Man , It is therefore necessary to be particular in selecting Men for this Service not only of Activity and Bodyly Strength but also of some Experience and approved Spirit.[73]

Although the British Army adapted to North America in other ways, there is some evidence to suggest that on a local level, the British Army used skirmishers. In 1778, the 71st Regiment of Foot was issued standing orders which indicated:

> If the Battalion is commanded to engage in a woods, thicket or country, one or more Sections will be detached in front of each Company with an Officer at the head of each who are immediately to occupy every Tree, Stump, Log, Bush, Rock, Cleft, Hedge, Wall, or in short, any kind of covering which can afford them tolerable shelter from the enemy…When the Signal for Action is given, the firings are immediately to commence on which Occasion every man shall take the most direct aim possible at the most Favourable Object in his front and without waiting for an Officer's orders with respect to times continue to load, present, and fire with the utmost alacrity, deliberation and accuracy 'til the firings are ordered to cease.[74]

---

72 Matthew Spring, *With Zeal and With Bayonets Only*, p.252.
73 Quoted in Raymond Henry Smythies, *Historical Records of the 40th (2nd Somersetshire) Regiment now 1st Battalion The Prince of Wales's Volunteers (South Lancashire Regiment)* (Devonport: Swiss, 1894) pp.549–551.
74 Orderly book of the 2nd Battalion of the 71st Regiment of Foot, MSSHM 617, Huntington Library, f.88, 'Standing Orders, 71st Regiment, June 30th 1778'.

Officers commanding these sections of what we might call skirmishers were ordered to:

> ... observe the same attention with regard to their particular place of their Sections in front of each Company and that their respective diversions shall not only be judiciously dispersed but that every Soldier shall hug their coverts in the most compleat manner possible for giving annoyance to the enemy and perfect security to themselves – If the troops are ordered to move in any direction they are to spring from tree to tree, Stump, Log, & etc with the utmost Agility & continue to fire, load and spring as they advance upon or retreat from the enemy. If the Point of War is beat, they are to rush upon the enemy with Charged Bayonets.[75]

In describing the Saratoga campaign, Lieutenant General John Burgoyne reported that a British column at Saratoga was preceded by, 'scouts and flankers,' but this could simply mean troops guarding the flankers of the column. Interestingly, he also states that, 'the picquets, which made the advanced guard of that column, were attacked in force, and obliged to give ground, but they soon rallied and were sustained.'[76] This language appears to share some similarities with the descriptions above.

The Battle Against Cavalry

Infantry officers who argued for rigidly controlling their soldiers stood on the firmest ground when the threat from enemy cavalry appeared. Fighting between bodies of infantry was much more common, but cavalry attacks against infantry were a feature of most major eighteenth-century battles. In this way, the fighting at Parma and Guastalla in the 1730s was an outlier, as was much of the fighting that will close this book during the American War of Independence. In these conflicts, cavalry charges against infantry occurred but were infrequent. Another example of its relative infrequency is the rarity in descriptions of this type of fighting that we have from infantrymen. Our enlisted infantrymen writers leave descriptions of infantry firefights, but rarely describe fighting against cavalry. Thus, we must turn to both theorists and officers, in addition to enlisted men themselves, for information regarding this type of combat.

No experience of battle was as terrifying for infantrymen as the sudden onset of cavalry. The French (and American) cavalry officer Augustin Mottin de la Balme commented: 'Indeed, where can we find men who will keep their cool in a terrible moment when they are facing down the charge of a body of well-led cavalry? Is there

---

75 Orderly book of the 2nd Battalion of the 71st Regiment of Foot, MSSHM 617, Huntington Library, f.88, 'Standing Orders, 71st Regiment, June 30th 1778'.
76 Anon., *A Brief Examination of the Plan and Conduct of the Northern Expedition in America in 1777* (London: T. Hookham, 1779), p.33.

any moment in war so forcefully destructive, except the explosion of mine? Will we find a single military man who is not in awe and terror of this moment?'[77]

So, infrequent or not, all infantry soldiers lived in terror of this event. De la Balme's comparison of charging to an explosive event matches well with the psychological terror of facing down charging enemies on horseback. Only soldiers who stood their ground had a reasonable chance of survival; those who fled would be cut down. Thus, officers lived in fear of soldiers who would break ranks in front of enemy cavalry. Count Turpin de Crissé commented:

> Great care must be taken to only fire in platoons, and when the cavalry is thirty paces away, if we allowed the enemy to approach closer, the soldier, who does not reason like the officer, is frightened by the horse and the rider… he would retreat, enabling the enemy to enter the battalion. If kept thirty paces away, the distance is still far enough that the infantrymen will not be intimidated, but is still close enough for the fire to have the necessary effect.[78]

Here, Turpin de Crissé asserts that soldiers, if not carefully managed, would break ranks and flee before enemy cavalry, since they do not 'reason like the officer.' In practice, soldiers often stayed in their ranks because of the belief that they would be annihilated in they fled. Authorities such as the Marqués de Santa Cruz argued that solid soldiers in order ranks provided the best means to survive the onset of cavalry. He argued, 'if your infantry detachment comes to rout that of enemy cavalry, it must only pursue in good order… because there would be the fear that the infantry would break their ranks'.[79] Here, Santa Cruz is not afraid of the enemy breaking through during a charge, but of the battalion becoming disorganized in the exuberance of pursuit.

Despite this, many infantrymen continued to believe that they held an advantage over cavalry, even when isolated or in small parties away from the protection of the battalion. In the little war of posts, American pensioner Benjamin Jones described a terrifying skirmish between his infantry and enemy cavalry:

> They had been out all night and while halting a gentleman rode up and asked where the commander of the scout or party was. Ensign Smith, who had the command said I am here[.] He said you have got 60 light horse in a quarter mile of you Smith said I care not for that. The man rode away. Smith ordered the men to ready, and you saw the light horse coming in sight. On full speed, the scouting party struck across the fields and the light followed and they formed into a hollow square. They formed around his party there being only

---

77 Augustin Mottin de la Balme, *Élémens de tactique pour la cavalerie* (Paris: Jombert and Ruault, 1776), p.94.
78 Lancelot Turpin de Crissé, *Essai sur L'art de la Guerre* (Paris: Jombert and Frault, 1754), pp.204.
79 Alvaro Navia Osorio, Marqués de Santa Cruz de Marcenado, *Reflexions militaires et politiques* (Paris: Rollin, 1736), vol.3, p.67.

27 of his party. The commanding officer of the horse told Smith if he would resign himself up he should be used like a prisoner or he would parol[e] him and his men Smith told him he should not do it the officer of the horse said if we have to fight and take you we shall cut you into pieces Smith said you must take us first the officer said he would give five minutes to surrender Smith said charge & be damned every man was ordered on his right knee & the b[reech] of his gun on the ground and Smith stood in the centre and told his men the first that gave back he would Cut his head off with his sword then one quarter of the horse charged on them and their horses were pricked and one of the horses was thrown and the rider fell over into the Hollow square and Smith put his foot to the horsemans sword and said I have got one I want some more charge again The Horse made another charge and were repulsed. & the third charge was made and repulsed again and our party took a prisoner and killed a horse and one of his own mens arm was broken and bayonet was broken and the horse rode off and formed. And our men all raised on their feet and rested and the prisoners sat in the centre. Smith then ordered one of the soldiers to take the commander off of the horse as he was parading his men The soldier drew up and shot him and he the officer fell dead. Then another officer took the command and he was ordered to be shot which was done – & the third took command and rode out from the horse and said to Smith if you will give up it shall be well if not we will send for 100 more horse and have you we will. Smith told him to send and be d[amne]d I want to manure the ground with the tories so that it should bear something after the war. Two of the horse were dispatched immediately. Smith ordered his men into rank & file at two paces distance in front & rear opposite to the spaces so as to fire through them. Smith ordered the front rank to begin a scattering fire on the right & to fire from the left & then the rear to do the same from left to right & every man to take good aim, which was done which drove the horse off there were but 24 left the field beside the two that had been sent away the rest were taken or killed or wounded.[80]

Jones's account was worth reproducing at length, as it gives a moment-by-moment account of fighting between infantry and cavalry, with detail rarely provided by sources from enlisted men. What does Jones's account indicate? First, these troops of Captain Bank's Company of David Waterbury's Provisional Connecticut Regiment do not seem to have been overwhelmed by the appearance of enemy cavalry, even though they were outnumbered over two to one. An initial unsuccessful attempt to flee ended with the isolated troops forming into hollow square, a standard response to isolation among infantrymen, whether facing cavalry or not. Second, after successfully receiving a number of charges, seemingly without firing, Ensign Smith responds by deploying his outnumbered infantrymen in extended order and firing. As opposed to holding his men in close order, Smith decided to form an alternating line, two ranks deep, in

---

80 NARA: M804, Benjamin Jones, Pension Application of Benjamin Jones, S13565. I would like to thank the 'If I Recollect Right: Rev[ultionary] War Pension Narratives' Project for first bringing this account to my attention.

extended order, for maximum fire effect. This seems to have had the intended effect. While Ensign Smith initially formed his men into a solid body to resist the enemy, he was not afraid to deploy his men in a looser formation in order to drive off the cavalry. Third, related to the matter of aimed fire, these troops, clearly armed with weapons that could fix a bayonet, picked off multiple enemy officers. When so ordered, troops without rifled weapons were more than capable of incapacitating enemy commanders. Infantry remained a more flexible and adaptable force, capable of a greater range of action, in many eighteenth-century battles. Multiple other pensioners attested similar events in the later American War of Independence.[81]

Indeed, Ensign Smith's reaction to the appearance of enemy cavalry matches well with the commentaries of the Marqués de Santa Cruz. Santa Cruz asserted that 'infantry parties sent against cavalry must be composed of good soldiers, commanded by the most valiant officers.'[82] He further recommended that 'infantry officers, in order to prepare their soldiers,' should mount a strong and healthy horse will a man of their choice, and have him charge an infantrymen.[83] The infantrymen selected as the target, awaited this advance with only a stick. 'Your men will see,' Santa Cruz elaborated, 'that by only waving the stick in the eyes of the horse, or by touching the horse on the head, the animal will swerve to the right or left.'[84] This would have the good effect of demonstrating that while their infantrymen might fear the horses of the cavalry, the horses also feared them. Santa Cruz concluded:

> From this the officers will take the opportunity to represent to the men that if a horse is frightened by a man who stands firm, having only stick in his hand, they will find that the efforts of cavalry will be useless against closed battalions, whose bayonets, musket balls, gleaming weapons, smoke, and noise of gunpowder make them even more capable of terrifying horses.[85]

Like so much eighteenth-century warfare, then, the battle against cavalry was primarily one of intimidation on both sides. Would the charging cavalry press home against unsteady infantry, or would steady infantrymen cause the horses to shy away? But what would happen if the infantrymen lost this contest? Christopher Duffy commented on this possibility at length:

> If the worst came to the worst, and the infantry were swamped by cavalry, the best hope of survival was to lie flat on the ground, hoping to stay out of reach of the sword blades, and get up again after the cavalry had passed on to other business. The *Bataillon-Garde* was overrun twice in this way at Kolin in 1757, yet was able to retire from the field in a semblance of order.[86]

---

81   NARA: M804, Pension Application of James Croft W20931; NARA: M804, Pension Application of Asa Jones S13568.
82   Santa Cruz, *Reflexions militaires et politiques*, vol.3, p.67.
83   Santa Cruz, *Reflexions militaires et politiques*, vol.3, p.68.
84   Santa Cruz, *Reflexions militaires et politiques*, vol.3, p.68.
85   Santa Cruz, *Reflexions militaires et politiques*, vol.3, p.68.
86   Christopher Duffy, *The Army of Frederick the Great* (2020), p.138.

The third rank of the Prussian regiment of Bevern tried to do the same at Kolin… but its fire had little effect and the troops were virtually wiped out. Lieutenant Prittwitz began to get up after a storm of horses had passed over him, but a veteran NCO shouted to him to lie down again. This was good advice, for a prone figure was out of reach of the cavalrymen's swords, and Prittwitz survived with only cuts and bruises.[87]

Once again, these examples demonstrate the flexibility of responses available to eighteenth-century infantrymen. When stoicism in the face of enemy cavalry failed, the men resorted to flexibility, even reprimanding their officers when they took actions the men deemed unsafe. Theorists who argued for the primacy of rigid control believed that never allowing battalions to fall into disorder would enable them to survive attack by enemy cavalry. But much like the willingness to engage in aimed firing and fight as skirmishers, men obeyed their officers' commands when those commands made sense, yet were not afraid to act on their own when their lives were at stake.

Conclusion

Between 1733 and 1783, officers considered and implemented their theories on infantry combat. Two key areas of debate and reform were the use of aimed fire and regular infantry troops fighting as skirmishers. As we have already seen at the Battles of Parma and Guastalla, officers realized the importance of skirmishers to delay an enemy advance and valued soldiers who could pick their own targets and engage at will. These ideas continued to hold importance throughout infantry warfare in the eighteenth century. Both of these concepts show that officers were willing to impart a level of trust to their men. Both in aiming their shots and fighting as skirmishing troops, line infantrymen were granted a level of initiative not usually allowed in the mythic automaton view. Both of these areas illustrate a moment in eighteenth-century tactics where officers were willing to relax a portion of their control on the battlefield, delegating tactical authority to their men.

The struggle against cavalry highlighted the importance of keeping men together in an ordered body, which could intimidate the horses of enemy cavalry into not bowling over the battalion. Even officers who expected some flexibility on the part of their men, such as the Marqués de Santa Cruz, were relative proponents of order in this situation. If that solidity failed, though, the men had a backup plan: throwing themselves to the ground, out of reach of the enemy sabres. Having done so, they looked out for their officers, who might be unfamiliar with the realities of this situation. In the ongoing battlefield negotiation between officers and men, soldiers were not passively waiting for their officers to delegate their authority. In many instances, these men would seize battlefield decision-making away from their officers, as they actively disobeyed them in combat.

---

87 Duffy, *The Military Experience in the Age of Reason*, p.216

7

Disobeying the Officers: Range, Intensity, and Duration of Eighteenth-Century Firefights

On 1 October 1756, Ulrich Bräker, a Swiss soldier in the Prussian Army, found himself in a tight spot. Historians disagree on whether Bräker was truly dragooned into the army as he claimed or volunteered of his own free will, but regardless, as the Prussian army marched to attack Austrians in defensive positions on a mountainside, Bräker felt distinctly queasy. He described the scene:

> We moved forward continuously. My courage deserted me. I would like to have been able to disappear into a hole in the ground, and I saw a similar fear on the faces of others, even those who had previously boasted of their courage. Empty flasks sailed through the air as soldiers finished them, and they said, 'courage for today, and perhaps no need of it tomorrow!' Now we advanced under artillery and took over a place in the first line of battle. By heaven! How the iron flew over our heads, knocking holes in the ground and kicking up stones and earth![1]

Prussian troops continued to press their advantage, chasing the Austrians back through the town of Lobositz, and eventually forced the Austrian army to withdraw. This phase of the Battle of Lobositz demonstrates incredible flexibility on the part of Prussian soldiers. A Prussian officer described the scene: 'The Itzenplitz Regiment and the following regiments of our left flank made an advance, the enemy right flank retired into the city, and they fired together with the Pandours out from the windows.'[2]

Far from being constrained by their officers and only envisioning combat occurring in the rigid lines of linear warfare, these soldiers continued to operate even after the cohesion of regimental organization had broken down. Bräker noted, 'the attack now began on Lobositz itself. I wasn't in the fore, but back in the rear a little way up among the vineyards. Many men as I have said, sprang down more nimbly than I from

---

1   Ulrich Bräker and Johann Heinrich Füssli, *Lebensgeschichte und natürliche Ebentheuer des Armen Mannes im Tockenburg* (Zürich: Füssli, 1789), p.148.
2   Geheimes Staats Archiv Preussischer Kulturbesitz (GStAPK): IV HA Rep.15A, Nr.681, Aus den K[öniglichen] Lager bey Pirna 3 Oct. 1756.

wall to wall, hastening to aide their comrades.'³ In Bräker's telling, soldiers fired independently, with some forces pressing the enemy with bayonet attacks, while others stayed on higher ground and fired at enemy troops. Unlike the Battle of Mollwitz 15 years earlier, Prussian troops in combat easily conceived of combat outside the orderly firing of platoons.⁴ The cost was heavy: a casualty report indicated that the Itzenplitz Regiment alone lost 243 men: 178 wounded and 65 men 'shot dead.'⁵

In aiming their fire and skirmishing, most eighteenth-century soldiers were happy to go along with their officers' orders. In contending directly with the enemy's main force, however, soldiers often disappointed officers who believed that the best way to engage the foe was to advance into close range without firing, or let the enemy do the same. To fully see the nature of eighteenth-century firefights, this chapter will explore their range, the tendency of soldiers to fire independently, ammunition usage, and rate of fire. Taken together, these factors provide a window into the nature of combat between opposing bodies of infantry during the period between 1733 and 1783.

Range of Firefights

In the eighteenth century, the term 'musket-shot' usually referred to a distance of around 300 yards. Commanders, concerned with the accuracy of their musketry, often performed tests in peacetime to discover an optimal range. During the era, such tests were conducted at between 500 yards and 80 yards, 200 yards and 100 yards, 250 yards and 80 yards, and 200 yards and 80 yards.⁶ The current author finds it strangely compelling that none of these tests felt the need to practice at ranges underneath 80 yards, perhaps implying that combat infrequently reached those close distances, or that accuracy under this distance was all but assured. Leaving aside this theoretical point, let us turn to what the soldiers of the era say in their writings: how close in range were eighteenth-century firefights?

Although there is no way to truly measure the surviving descriptions of range scientifically, we can perhaps arrive at a few conclusions regarding the range of eighteenth-century firefights. In the first category, there are skirmishes and premature fires by inexperienced troops. These preliminary skirmishes often occurred at 300 yards, or even a greater distance. As a result, they were not very deadly. During skirmishes in 1759 and 1760, French troops and their native allies opened fire on the British at a range of 300 yards, which seemed quite normal to

---

3   Bräker and Füssli, *Lebensgeschichte*, p.155.
4   Duffy, *The Military Experience in the Age of Reason*, p.212.
5   GStAPK: IV HA Rep.15A, Nr.681, Verlüst der Preüβischen Armee in der glorieusen Bataille mit der Käyserlich-Königl[ichen] Arme bey Lowositz in Böhmen.
6   Duffy, *The Military Experience in the Age of Reason*, p.207; Duffy, *The Army of Frederick the Great* (2020), p.146; Ludwig Matthias von Lossow, *Denkwürdigkeiten zur Charakteristik der preussischen Armee, unter dem grossen König Friedrich dem Zweiten. Aus dem Nachlasse eines alten preussischen Offiziers* (Glogau: Carl Heymann, 1826), pp.260–261, 275.

the participants.⁷ During a skirmish on Staten Island in June of 1777, preliminary skirmishing began at about 300 yards.⁸ At the Battle of Mollwitz in 1741, the Prussian infantry, inexperienced in real combat, opened fire at the considerable range of 600–800 yards.⁹

In Europe during large-scale and determined combat, battlelines could draw much closer. A good 'average' range for combat during the Seven Years War appears to be between 200 yards and 100 yards. At Guastalla in 1734, the Austrian advance broke down into uncontrolled firing at 100 yards.¹⁰ At Dettingen in 1743, a British colonel, Scipio Duoure, wrote, 'at last, the British begged of their officers to let them fire, which (when the enemy were about sixty yards from us) they did.'¹¹ The role of the troops, begging the officers to let them fire, highlights that enlisted men viewed 60 yards as extremely close range. During a skirmish near Cuneo in Northern Italy in September of 1747, an officer observing the French and Austrians simply noted, 'The French fired at so great a distance, that very few were killed or wounded.'¹² At the Battle of Leuthen in 1757, Ernst von Barsewisch recalled, 'As soon as we had cleared the forest, we approached the enemy's second line of battle at a distance of 200 paces [150 yards], which was preparing to march against us. Now... our officers ordered, 'Fire! Fire!'¹³ Later in the war, Barsewisch recalled being fired on by Croats at a similar distance.¹⁴ The battle lines at Prague in 1757 appear to have been around 150 yards (200 paces) apart.¹⁵ At Hochkirch in 1758, Johann von Archenholz recalled that the Austrians opened fire at 'a few hundred paces.'¹⁶ During a skirmish near Prenzlow in October of 1760, hostile forces approached to within 200 paces (150 yards) of one another.¹⁷ At the Battle of Vellinghausen in 1761, official reports indicate that battle lines were 150 paces (100 yards) apart, and indicate that this was uncomfortably close, noting, 'the troops on both sides stood scarcely 150 paces apart at many points, and... shot at one another to great effect.'¹⁸ By contrast, when

---

7   John Knox, *An historical journal of the campaigns in North America for the years 1757, 1758, 1759, and 1760* (London: 1769), vol.1, p.305; vol.2, p.274.
8   Nicholas Cresswell, Harold B. Gill Jr, and George M. Curtis III, *A Man Apart: The Journal of Nicholas Cresswell, 1774–1781* (Blue Ridge: Lexington Book, 2009), p.170
9   G. L. Valori, 'Observations sur le service militare du roi de Prusse,' in Reinhold Koser, *Forschungen zur brandenburgischen und Preussischen Geschichte* (Berlin: Duncker & Humblot, 1894), vol.7, p.308.
10 Historical Manuscripts Commission, *Report on Manuscripts in Various Collection*, vol.8, p.408.
11 Edward Arthur Howard Webb, *History of the 12th (The Suffolk) Regiment, 1685–1913* (London: Spottiswoode & Co, 1914), p.64.
12 TNA: SP 87/23, f.382, Lt. General Wentworth to the Duke of Newcastle, Demont[e], September 22nd 1747.
13 Ernst Friedrich Rudolf von Barsewisch, *Meine Kriegs-Erlebnisse während des Seibenjährigen Krieges, 1757–1763* (Berlin: von Warnsdorff, 1863), p.36.
14 Barsewisch, *Meine Kriegs-Erlebnisse*, p.170.
15 Johann Wilhelm von Archenholz and Max von Duvernoy, *Geschichte des Seibenjährigen Krieges in Deutschland*, (Leipzig: Amelangs, 1911), p.52.
16 Archenholz and Duvernoy, *Geschichte des Seibenjährigen Krieges*, p.187.
17 Anon., *Sammlung ungedruckter Nachrichten so die Geschichte der Feldzüge der Preußen von 1740 bis 1779* (Dresden: Walther, 1785), vol.5, p.358.
18 Christian Heinrich Philip Edler von Westphalen, *Geschichte der Feldzüge des Herzogs Ferdinand von Braunschweig-Lüneberg* (Berlin: E. S. Mittler & Sohn, 1872), vol.5, p.623.

the battle lines closed to 80 yards at the Battle of Minden, the contest was no longer in doubt, and the French began to retreat.[19] The Prussians at Kesselsdorf held their fire to a very close range of approximately 50 yards.[20]

In North America, infantry firefights followed a similar mould. Although the French and Indian War saw few large-scale field battles, a few sources describe infantry firefights. At the Battle of the Plains of Abraham, the French opened fire at regular European distance – 130 yards – while the British held their fire to 40 yards.[21] During fighting in the Caribbean in 1759, Francis Downman noted that the enemy moved away from the British, 'keeping always 200 yards in our front.'[22] During the American War of Independence, firefights continued to follow a European pattern, at least with regards to range. During the flank attack at the Battle of Brandywine, both the British and Americans gave fire at 150 yards, and then the British immediately charged at the run with their bayonets.[23] At Long Island, some of the army commenced firing at the hopeless distance of 300 yards.[24] Others retreated from the British advance at 80 or 100 yards distance.[25] At Trenton, the Hessians were convinced to surrender because Continental troops advanced to within 20 yards with loaded muskets.[26] At Princeton, some Americans managed to approach within 25 yards of the enemy.[27] At Brandywine, Patrick Ferguson recalled being fired on at 80 yards.[28] Other Continentals at the same battle managed to hold their fire to the more impressive range of 40 yards.[29] Another officer insisted the 'common distance' between the two armies at this fight was 50 yards.[30] In the skirmish at Matson's Ford in December of 1777, the American's fired at 200 yards.[31] In a skirmish on 25 April 1779, Continental troops fired at 20 yards.[32] At the Battle of Blanford in April

---

19 Westphalen, *Geschichte der Feldzüge des Herzogs Ferdinand*, vol.3, p.486.
20 Anonymous, *Sammlung ungedruckter Nachrichten*, vol.1, pp.430–431.
21 Knox, *An historical journal of the campaigns in North America*, vol.2, p.70.
22 Francis Downman and Francis Arthur Whinyates, *The Services of Lieut.-Colonel Francis Downman, R.A. in France, North America, and the West Indies between the Years 1758 and 1784* (Woolwich: Royal Artillery Institution, 1898), p.8.
23 Martin Hunter, *The Journal of General Sir Martin Hunter and some letters of his Wife Lady Hunter* (Edinburgh: The Edinburgh Press, 1894), p.29
24 Samuel Richards, *Diary of Samuel Richards* (Philadelphia: Leeds and Biddle, 1909), p.36.
25 Nathanael Greene and Richard Showman, *The Papers of General Nathanael Greene* (Chapel Hill: University of North Carolina Press, 1976), vol.1, pp.300, 307.
26 Testimony of Ensign Friedrich von Zeuge, *Hessian Documents of the American Revolution* (Boston: G.K. Hall, 1989), p.M.L.375, microform.
27 'Diary of Lt. James McMichael,' *Pennsylvania Magazine of History and Biography*, vol.16, no.2 (1892), p.141.
28 University of Edinburgh Archives, Laing Manuscripts, Patrick Ferguson to George Ferguson, 8 October 1777.
29 Friedrich Ernst von Münchhausen, *At General Howe's Side* (Monmouth Beach: Phillip Frenau Press, 1974), pp.31–32.
30 'Diary of Lt. James McMichael,' p.150.
31 'Memoirs of Brigadier General John Lacey,' *Pennsylvania Magazine of History and Biography*, vol.16, no.2 (1892), p.108.
32 LOC: Washington Papers, Lt. Thomas McClellan, 'Journal of my Route to Swegatia', in Phillip Schuyler to George Washington, 10 May 1779.

of 1780, American troops managed to fire at 60 yards.[33] At Camden in August of 1780, officers under Horatio Gates explicitly ordered their men to ignore Gates's instruction to fire at six paces and fired at a range of 70 or 80 yards.[34] At Green Spring in 1781, American troops fired 100 yards according to Captain John Davis.[35] Occasionally, American troops were able to hold their fire until close range, usually with disastrous results for the British opposing them. Impressively, at Cowpens, Continentals fired at 10 to 15 yards.[36] In the fight at the second line at Guilford Courthouse, American forces fired at around 20 yards.[37] This achievement was all the more impressive, as the British opened fire on the Americans, who did not reply until the British closed to a closer range.

The French attack on St. Lucia in December of 1778 was conducted 'at a distance not less than 280 yards.'[38] At the Battle of Camden, John Robert Shaw describes an infantry firefight at 100 yards.[39] At Guilford Courthouse, Roger Lamb reports that the American line opened fire at 140 yards.[40] Berthold Koch, a sergeant with the von Bose Regiment, recalled receiving fire at 100 yards from the American line.[41] Troops were often ordered to hold their fire until very close to the enemy line, but appear to have found it difficult to follow this directive. At the Battle of Germantown in 1777, Joseph Plump Martin describes this in his humorous way:

> Our brigade moved off to the right into the fields. We saw a body of the enemy drawn up behind a rail fence on our right; we immediately formed in line and advanced upon them. Our orders were not to fire till we could see the buttons upon their clothes, but they were so coy that they would not give us an opportunity to be so curious, for they hid their clothes in fire and smoke before we had either time or leisure to examine their buttons.[42]

Here, we find a unit ordered to hold fire until close to the enemy, but because the enemy began returning fire at a long range, a longer-ranged firefight developed. When troops did approach (or fire) at ranges closer than 80 yards, it was often because a bayonet attack was underway. The Swedish army in the Great Northern War, and the British Army in the American War of Independence, both made quick-moving

---

33   NARA: M804, William Miner, Pension Application of William Miner S11070
34   NARA: M804, Thomas Moody, Pension application of Thomas Moody W25732
35   NAM: 1991-09-117, John Davis Papers, Journal of Captain John Davis.
36   Thomas Anderson, 'Journal of Lieutenant Thomas Anderson of the Delaware Regiment, 1780-1782,' *Historical Magazine*, vol.2, no.1 (1867), pp.208–209
37   NARA: M804, Jacob Salmons, Pension Application of Jacob Salmons R9157
38   Downman, *The Services of Lt. Colonel Francis Downman*, p.98.
39   John Robert Shaw, Oressa M. Teagarden, and Jeanne L. Crabtree, *John Robert Shaw: An Autobiography of Thirty Years, 1777–1807* (Athens: Ohio University Press, 1992), p.31.
40   Lamb, *An Original and Authentic Journal of Occurrences during the Late American War*, p.350.
41   Berthold Koch and Bruce E. Burgoyne, *The Battle of Guilford Courthouse and the Siege and Surrender at Yorktown* (Greensboro: Guilford Courthouse National Park, 2014), p.7.
42   Joseph Plumb Martin, *A Narrative of some Adventures, Dangers, and Sufferings of a Revolutionary Soldier,* (Hallowell: Glazier, Masters & Co, 1830), p.53.

assaults supported by one close-range volley a standard of their tactical repertoire. Nicholas Creswell describes this type of attack on Staten Island in 1777:

> When [the two sides] were about 100 yards from each other, both parties fired, but I did not see any fall. They still advanced to the distance of 40 yards or less and fired again. I then saw a great number fall on both sides. Our people rushed upon them with their Bayonets and the others took to their heels. I heard one of them call out *murder* lustily, this [would have been] laughable if the consequence was not serious. A Fresh party immediately fired upon our people but was dispersed and pursued into the woods by a company of the 15th Regmt.[43]

Here, Nicholas Creswell describes the archetypal bayonet attack as practiced by the British Army in the American War of Independence. If troops moved into 50 yards or closer, it was not to have an extended firefight, but because one or other intended to make an attack with bayonets, or was in the process of doing so. Some British commanders may have preferred to try and close the distance before firing. Drill manuals, books offering advice to officers, etc., often instructed soldiers to reserve their fire until 30 or 50 yards from the enemy.[44] Often, the commanders judged their tactics on a case-by-case basis. At the Prussian 1785 Review, British general Earl Cornwallis criticized Prussians for advancing too close to their mock enemies before opening fire.[45] While a good idea in theory, advancing into such close range under a heavy fire proved difficult to do. Notably, the British seem to have achieved this at the Battles of the Plains of Abraham and Le Belle Familie (1759), but examples of this kind are rare. On average, troops seem to have fired at ranges of 100 yards or longer.

Independent Firing

In the eighteenth century, commanders prized orderly and disciplined fire. The British and Prussian armies were famous for their level of discipline and firepower. Both of these armies were popularly known at the time and remembered today for firing by platoons.[46] Soldiers and junior officers regularly noted when troops were able to achieve this level of discipline and control on the battlefield. At battlefields as diverse as Mollwitz, Dettingen, Prague, the Sullivan Campaign, and Reichenbach, soldiers maintained complicated firings by platoons, divisions, and battalions.[47] We

---

43　Cresswell et al, *A Man Apart*, p.170
44　Blackmore, *Destructive and Formidable*, pp.136–137.
45　Charles Cornwallis and Charles Ross, *Correspondence of Charles, First Marquis Cornwallis* (London: John Murray, 1859), vol.1, p.211.
46　See Chapter Two for an extended discussion of platoon firing.
47　Duffy, *Military Experience in the Age of Reason*, pp.212–213; John Sullivan and Frederick Cook, *Journals of the Military Expedition of Major General John Sullivan against the Six Nations of Indians* (Auburn: Knapp, Peck & Thompson, 1887), p.232; Anon., *Fernere Schreiben eines Holländischen Volontairs* (Berlin: Publisher Unknown, 1757), vol.8,

should take these reports seriously – these armies were sometimes able to achieve fire discipline.

Despite this, troops involuntarily used another type of fire, something that in modern English we might call 'firing at will.' In the eighteenth century, the British sometimes called this an 'irregular,' 'straggling,' or 'running fire;' the French a, '*feu de billebaude;*' and German sources often refer to it as '*Plackerfeuer*' or sometimes '*Batalillenfeuer.*' In practice, this simply meant that troops loaded and fired as quickly as they could, often without orders from officers. It is important to carefully interrogate sources referring to 'running fire,' as at times this can mean a quick but orderly fire by division. Eighteenth-century military theorists often frowned upon this type of firing, as it was perhaps less effective than controlled firing by platoon or rank. Despite the censure of military theorists, we can observe this type of firing in a number of sources. It was used on many eighteenth-century battlefields and across most armies. The British and Prussian armies took steps to mitigate firing at will and it may have occurred less frequently, but sources still report its use in those armies. Perhaps the most famous (and controversial) description of this type of firing comes from Lieutenant Colonel Russell of the British Guards at the Battle of Dettingen in 1743:

> That the Austrians behaved well also is true; that except one of their battalions which fired only once by platoons, they all fired as irregular as we did; that the English infantry behaved like heroes, and as they were the major part in the action to them the honor of the day is due; that they were under no command by way of Hide Park firing, but that the whole three ranks made a running fire of their own accord, and at the same time with great judgement and skill, stooping all as low as they could, making almost every ball take place, is true, that the enemy, when expecting our fire, dropped down, which our men perceiving, waited till they got up before they would fire as a confirmation of their coolness as well as bravery, is very certain; that the French fired in the same manner, I mean like running fires, without waiting for words of command, and that Lord Stair did often say he had seen many a battle and never saw the infantry engage in any other manner is as true.[48]

As you might expect, this statement has generated some controversy. David Blackmore has called the eyewitness's credibility into account and has suggested that Lord Stair's comment may only apply to the French infantry.[49] Russell was two miles away from the battle, but wrote this letter two months after the battle, and would have been able to discuss the events with officers closer to the scene. The state-

---

p.27; Charles Immanuel de Warnery, *Campagnes de Frédéric II, roi de Prusse, de 1756 à 1762* (1788), p.48; Blackmore, *Destructive and Formidable*, p.107; Anon., *Das Treffen bei Reichenbach in Schlesien* (1762), pp.12–13.

48 Historical Manuscripts Commission, *Report on the Manuscripts of Mrs. Frankland-Russell-Astley* (London: Mackie & Co, 1900), p.278.

49 Blackmore, *Destructive and Formidable*, p.107.

ment 'Lord Stair did often say' implies that this topic was a matter of conversation, at least among officers, after Dettingen. Even if we assume that Stair's statement only applies to the French, what a statement! Another officer reported on the same battle reported that 'the British infantry fired not by platoons (for there was no governing them) but with perpetual volleys from right to left, loading almost as fast as they fired without ceasing so that the French were forced to retreat.'[50]

If this practice happened at Dettingen, where else can we observe it in the eighteenth century? As we have seen, the Austrian troops at Guastalla fired uncontrollably without regard for the threats of their officers.[51] Likewise, as the first chapter noted, at Parma during the same 1734 campaign, a French officer recalled, 'we had given our men the freedom to fire at their pleasure.'[52] At Gross-Jägersdorf in 1757, a Russian officer recalled, 'the musketry from our side was so intense that every soldier fired as soon as he had loaded, without waiting for the order of his officer, firing as fast as possible.'[53] Another Russian officer recalled the scene: 'The smoke from the firing was so thick that we could no longer see both armies… both front lines were quite close and fired at one another without pausing… a ceaseless firing had begun on both sides.'[54] Certainly, this practice was embraced by the French. In the 1750s, the comte de Chabot argued that allowing soldiers to fire at will was superior to other systems of fire: 'The French leave each man the will and power to direct his fire, and all this fire takes good effect… this is the great advantage of French fire.'[55]

Chabot put forth eight reasons why the independent firing (what he called, 'the French firing') was superior to the regulated firing of volleys practiced by other nations. First, the men had the freedom to choose their targets at distance, when they might need to aim higher or lower in order to hit their target.[56] Second, it allowed the men to aim their fire horizontally, even if the front of the battalion was angled in a different direction.[57] Third, it vastly increased the speed of fire by allowing recruits and veterans to load at their own speed, repair malfunctioning or jammed weapons, and continue firing even if their officers had been hit.[58] Fourth, it gave the men the time to properly load and seat the charges in their weapons, resulting in more accurate fire.[59] Fifth, it gave the soldier a better idea if his weapon had actually fired or not, preventing him from uselessly reloading an already charged weapon.[60] Sixth, it

---

50 Webb, *History of the 12th (The Suffolk) Regiment*, p.64.
51 Historical Manuscripts Commission, *Report on Manuscripts in Various Collections*, vol.8, p.408.
52 D'Espié, *Memoirs de la guerre d'Italie*, pp.166–167.
53 Hans Heinrich von Weymarn, 'Ueber den ersten Feldzug des Russischen Kriegsheeres Gegen die Preußen im Jahr 1757', *Neue Nordische Miscellaneen*, vol.7 (1794), p.198.
54 Andrei Bolotov, Жизнь и приключенія Андрея Болотова (St. Petersburg: Golvina, 1870), pp.526, 534.
55 Comte de Chabot, *Réflexions critiques sur les differens systêmes de tactique de Folard* (Berlin and Paris: Neaulme and Giffrat, 1756), p.9.
56 Chabot, *Réflexions critiques*, p.9.
57 Chabot, *Réflexions critiques*, pp.11–12.
58 Chabot, *Réflexions critiques*, pp.12–14.
59 Chabot, *Réflexions critiques*, p.14.
60 Chabot, *Réflexions critiques*, pp.15–16.

prevented several men from deliberately hitting the same target in a volley.[61] Seventh, by offering a continually rolling fire, it did not psychologically give the advancing enemy a break where they could advance without worry during the enemy's reloading.[62] Eighth, it did not require the officers to be continually shouting at the men, so that if an emergency occurred, the officers would be more readily heard.[63]

Comte Turpin de Crissé, writing in 1769, called the *feu de billebaude* 'the best of all fires,' and advocated that officers facilitate the fire by having men in the rear ranks load, while the men in the front ranks fired.[64] Austrian Veteran Jacob de Cogniazzo discussed the use and disadvantages of an 'unregulated fire.'[65] Tobias Smollet's military history, published in 1785, refers to straggling and irregular fire in reference to the British and French infantry in 1758, but since he was not an eyewitness, we should treat these accounts with care.[66] John Knox reports that the French used 'a galling though irregular fire' at Quebec in 1759.[67]

The Comte de Guibert, discussed the varieties of infantry fire in his 1772 essay on tactics. He concluded:

> Finally, the *billedbaude* is the only one that must be used in a battle of musketry; by the time two volleys are given and received, there is no effort of discipline that could hinder a complicated and regulated fire from degenerating into voluntary fire. This fire is the liveliest and deadliest of all; it excites the head of the soldier, it inures him to danger [and] it particularly corresponds to French vivacity and skill; the essential [task] is only to accustom the soldier to stopping at the signal and to maintaining silence. This was once regarded as impossible; today it will be easily achieved. In a battle in the Seven Years War, I have seen a regiment execute this fire under that of the enemy, beginning and ending at the drum signal. This regiment, despite being levied only four years prior, always fought with the same discipline and the same valor…[68]

The Prussians fretted endlessly about their inability to perfect platoon fire, even though they were more successful in achieving this end than other armies. Charles Immanuel de Warnery, writing in 1782, summarized some of the difficulties in firing with platoons:

---

61 Chabot, *Réflexions critiques*, p.16.
62 Chabot, *Réflexions critiques*, pp.16–17.
63 Chabot, *Réflexions critiques*, pp.17–19.
64 Lancelot Turpin de Crissé, *Commentaires Sur Les Mémoires De Montecuculi* (Paris: Lacombe, 1769), pp.179–180.
65 Jacob de Cogniazzo, *Geständnisse eines Oestreichischen Veterans* (Breslau: Gottlieb Löwe,1788), vol.1, p.169.
66 Tobais Smollet, *The History of England from the Revolution to the Death of George the Second* (London: Cadell, 1785), vol.4, pp.254, 260.
67 Knox, *An historical journal of the campaigns in North America*, vol.2, p.128.
68 Guibert and Abel (trans.), *Guibert's General Essay on Tactics*, p.78. He was describing the Royal Deux-Ponts Regiment at Vellinghausen.

The whole world seems to cry out against our platoon fire, since on the battlefield, we can only seem to do it twice. I agree in part, as I have written elsewhere, and I believe that we could perfect this system with better principles than we have now. The first vice is the size of our platoons: they are much too big, their frontage is too wide. How can we pretend that the officer who commands the platoon, standing on the right can see his command, much less be heard when the [more senior] two officers 50 paces to the rear can barely hear themselves? The noise of the artillery, musketry, and shouts of other platoon's officers, the cries of the wounded, blinding and suffocating smoke, the distance from where the [platoon] commander stands to the left of the platoon, all conspires to hide the officer from the platoon. He could not command them even if he had a voice of thunder. These are the primary obstacles which prevent us from firing by platoons, as the system currently exists. It is amazing that no one has tried to fix this. We should also carefully examine initial cause of the disorder which gets into the infantry as soon as it has started to fire. Officers agree that this is quite normal, [once the shooting starts,] they are almost no longer masters of their soldiers.[69]

In his 1790 handbook for officers, future Napoleonic theorist Johann von Scharnhorst entitles one of his chapter headings 'The Plackerfeuer which must be Avoided.'[70] He complains about firing at will for some time, giving detailed descriptions of why it is disadvantageous: 'The worst of all these is that a certain order in the battalion has been generally lost during this fire, and that the officers who have lost control of their command can only restore their attention with great difficulty, and often not at all.'[71] The extensiveness of his complaints indicate that he was not merely talking from a theoretical standpoint but from first-hand experience. Other Prussian military authorities also addressed the problem of troops involuntarily firing at will.

In 1798, Georg Heinrich von Berenhorst reflected on the art of war. He had served on Frederick II of Prussia's staff during the Seven Years War and would become an important military figure in the history of the Napoleonic Era. When describing infantry firing in battle, he noted:

> You begin firing by a salvo, or perhaps firing by platoons for two or three shots. Then a general blazing away follows: the usual rolling fire where everyone fires as quickly as they are loaded. Ranks and files become mixed, those in the first rank could not kneel even if they desired it, and the officers, from the lower ranks to the generals can do nothing more with this mass, until it finally begins moving forwards or backwards.[72]

---

69 Charles Immanuel de Warnery, *Remarques sur plusieurs auteurs militaires et autres* (Lublin: Staroludzki, 1782), pp.69–70.
70 Gerhard Johann von Scharnhorst, *Handbook für Officiere* (Hannover: Helwing, 1790), vol.3, p.276.
71 Scharnhorst, *Handbook für Officiere*, vol.3, p.277.
72 Georg Heinrich von Berenhorst, *Betrachtungen über die Kriegskunst* (Leipzig: Fleischer, 1798), vol.1, p.255.

This type of firefight is perhaps the kind of combat Ulrich Bräker was referring to when he recalled of the Battle of Lobositz in 1756. A soldier with the Itzenplitz Regiment, Bräker, recalled that in great heat and headiness, 'I fired away nearly all my sixty rounds. My musket became so warm that I had to carry it by the sling.'[73] A Prussian report from the Battle at Soor in 1745 indicated, 'In the meantime, our infantry had to endure a strong fire from the small arms [of the enemy]. Our battalions began to fire without orders, the enemy withstood this and continued his fire, which brought disorder into our lines.'[74] Johan Gottfried Hoyer, who would rise to prominence as a Prussian *generalmajor* of the Napoleonic Era, set out to describe military history between 1750 and 1799. In describing infantry fire-tactics, he describes firing by platoons and other various forms of complex firing in the eighteenth-century. Near the end of his description, he commented:

> In fact, all these types of firing were practiced in peacetime on the drill-square, but soldiers hardly used them in serious combat. Once there, everything was abandoned for running fire [plackerfeuer], that is, everyone loaded and shot for himself as fast as he could. This is highly embarrassing, as after one hundred years of practice, we cannot bring common soldiers under control, and build a robotic shooting-machine. In the heat and confusion of battle, the instrument is only set in motion by the artist's finger. Some exceptions [to the general rule of running fire], which may be found among the Prussian troops, and only with them alone, have been made possible through their ceaseless practice. They can prove nothing against the universality of the idea shared here.[75]

In North America, there are examples of this type of firefight as well, though during the American War of Independence, the British used bayonet attacks and volley fire. This swift-moving sort of attack sometimes prevented a general breakdown of control. On the other hand, the forces of the young United States appear to have used this tactic frequently.[76] It occasionally happened to the British as well, as British officer Thomas Anburey describes it in his letters from Burgoyne's campaign:

> In this action, I found that all manual exercise is but an ornament, and the only object of importance it can boast of was that of loading, firing, and charging with bayonets: as to the form, the soldiers should be instructed in the best and most expeditious method. Here I cannot help observing to you, whether it proceeded from an idea of self-preservation, or natural instinct, but the soldiers greatly improved the mode they were taught in, as

---

73 Ulrich Bräker and Johann Heinrich Füssli, *Lebensgeschichte und natürliche Ebentheuer des Armen Mannes im Tockenburg* (Zürich: Füssli, 1789), p.150.
74 Anon., *Sammlung ungedruckter Nachrichten*, vol.1, p.359.
75 Johann Gottfried von Hoyer, *Geschichte der Kriegskunst* (Göttingen: Rosenbusch, 1799), pp.102–103.
76 John Graves Simcoe, *Simcoe's Military Journal* (New York: Bartlett & Welford, 1844), pp.45, 146; Cook, *Journals of the Military Expedition of Major General John Sullivan*, pp.92, 95.

to expedition, for as soon as they primed their pieces, and put the cartridge into the barrel, instead of ramming it down with their rods, they struck the butt end of their piece upon the ground, and bringing it to the present, fired it off. The confusion of a man's ideas during the time of action, brave as he may be, is undoubtedly great...[77]

This phenomenon, often called 'tap-loading,' quickened the rate of fire but had the potential to greatly reduced muzzle-velocity. It is possible that the Austrians engaged in this practice at the Battle of Mollwitz in 1741.[78] If soldiers fired at longer ranges, and fired quickly, at times without regard for their officers' orders, how much ammunition did they use?

Ammunition Usage

What can non-data-driven sources tell us about the frequency of running out of ammunition? We must also carefully interrogate these sources, as running out of ammunition was often used by eighteenth-century soldiers as an excuse for failure, or in order to clear their reputations of any poor conduct in battle. With that being said, eighteenth-century soldiers ran out of ammunition quite frequently. Ammunition shortages plagued many major armies on the battlefield. In the era of Frederick II of Prussia, the Prussian Army developed new strategies for the rapid consumption of ammunition. At the Battle of Mollwitz in 1741, the Prussian troops quickly fired away their 30 issued rounds and attempted to gather ammunition from wounded men nearby. After the battle, the standard ammunition load in the Prussian Army was increased from 30 to 60 rounds, but even this proved insufficient.[79]

Two independent sources affirm that the Prussian infantry used all (or almost all) of their ammunition while attacking the Lobosch Hill at the Battle of Lobositz in 1756. An individual soldier reported firing away nearly his entire ammunition load.[80] The personal secretary of the Duke of Bevern, Herr Kirstenmacher, who observed his master at the battle, recorded the following:

> The greatest difficulties had to be overcome in order to dislodge the enemy. We were under a small arms fire which lasted for five hours at an unimaginable intensity. Our lads shot away all their cartridges, and those of their dead and wounded comrades. Now we had reached a crisis point, since the

---

77   Thomas Anburey, *Travels through the Interior Parts of America* (London: William Lane, 1789), vol.1, p.333.
78   Anon., *Denkwürdiges Leben und Thaten Johann Daniels von Menzel* (Frankfurt and Leipzig: 1743), p.80.
79   Anon., *Neue Militärische Zeitschrift* (Vienna: Anton Strauss, 1813), vol.2, Issue 19, p.21.
80   Ulrich Bräker and Johann Heinrich Füssli, *Lebensgeschichte und natürliche Ebentheuer des Armen Mannes im Tockenburg* (Zürich: Füssli, 1789), p.150.

enemy continued to fire heavily on our men, who did not flinch and could now not lift a finger to fire themselves.[81]

Despite the obvious melodrama in Kirstenmacher's telling, he clearly indicates the feeling of helplessness running out of ammunition could impart to soldiers. It is quite possible that these soldiers fired away 90 cartridges, rather than 60, as sources describe taking 30 rounds from unengaged units.[82] At the Battle of Leuthen in 1757, both sides replenished their rounds from ammunition carts, and it appears that some Prussian infantry may have fired over 180 rounds.[83] During the Battle of Zorndorf in 1759, many Prussians were wounded by the buckshot ammunition that the Russian infantry employed at that battle. At the Battle of Hochkirch in 1758, both the Prussian and Austrian infantry appear to have run short of ammunition.[84] In the Prussian case, some soldiers fired over 120 rounds.[85]

At the Battle of Torgau in 1760, multiple sources in the Austrian army reported that the primary reason for failure was a lack of infantry and artillery ammunition. Franz Moritz von Lacy reported, 'Finally, everyone was agreed that there was no more ammunition for either the artillery or the infantry [by the end of the battle.]'[86] Austrian veteran Jacob Cogniazzo reported that in this case, a previously useful innovation – filling drums with ammunition and employing drummers as ammunition runners – had failed the Austrians:

> ...the lack of ammunition, a defect which should never be found in a purposeful institution, and is always a sign of irresponsibility. But we had experienced it before, particularly at the Battle of Breslau. Sufficient ammunition was not brought forward, by way of the drummers and their drums, because they had to haul the ammunition from a very great distance, losing a great deal of time. This put the loading troops at a great disadvantage.[87]

When fighting in Continental Europe, the British often relied on firepower. A Dutch officer observed that at the Battle of Fontenoy the British fired all their cartridges, perhaps 20–36 per man.[88] During the Seven Years War, British infantry began to carry 30 or 60 rounds per man, rather the regulation 24. Yet more ammunition was ready-made in specialized wagons following the army.[89] In the Seven Years War in North America, British ammunition allocation seems to have fluctuated between

---

81 Curt Jany, *Urkundliche Beiträge und Forschungen zur Geschichte des preussischen Heeres* (Berlin: E.S. Mittler, 1901), vol.1, pp.9–10
82 Jany, *Urkundliche Beiträge*, vol.1, p.3.
83 Duffy *The Army of Frederick the Great*, (2020), p.302; Christopher Duffy, *Prussia's Glory: Rossbach and Leuthen 1757* (Chicago: Emperor's Press, 2003), p.158.
84 Duffy, *By Force of Arms: The Austrian Army in the Seven Years War Volume 2*, pp.142–144.
85 Barsewisch, *Meine Kriegs-Erlebnisse*, p.77.
86 Quoted in Duffy, *By Force of Arms*, p.300.
87 Jacob de Cogniazzo, *Geständnisse eines Oestreichischen Veterans* (Breslau: Gottlieb Löwe,1788), vol.3, p.298.
88 Quoted in Blackmore, *Destructive and Formidable*, p.136
89 Todd, *The Journal of Corporal Todd*, p.164.

36 and 70 cartridges per man. Despite the fluctuation, three extra flints appear to have been the standard issue.⁹⁰ In July of 1759, Townshend's Brigade received the following order: 'The light infantry of the army are to have their bayonets, as the want of ammunition may sometimes be supplied with that weapon: and, because no man should leave his post, under pretense that all his cartridges are fired, *in most attacks by night, it must be remembered, that bayonets are preferable to fire.*'⁹¹

The emphasis is present in the original. It is intriguing that the order commands troops to stay at their posts when out of ammunition. This idea – that soldiers with empty cartridge pouches could return to the rear – also confronted the American army at the Battle of Germantown in 1777.

In the American War of Independence, there are also numerous examples of soldiers running short on ammunition. Despite the British Army's clear preference for an aggressive moment, British soldiers found themselves drawn into heavy firefights. One of the clearest examples of replenishing ammunition from fellow soldiers comes from Sergeant Thomas Sullivan of the 49th Regiment of Foot. During the Forage War in early 1777, Sullivan reports that during a skirmish, 'Major Dilkes with [100 grenadiers] engaged them with two field pieces, and kept a continual fire up, until they expended all their ammunition at a rate of sixty rounds per man. Then they retreated to the second party of Grenadiers from whom they got more ammunition.'⁹²

At the Battle of Freeman's Farm, it is possible that the British infantry regiments were able to access 100 rounds. This was the ammunition allotment set as a standard for the army by Guy Carleton in April of 1777: '...every soldier in the army should always be provided with 100 cartridges, of which the man should have thirty in his cartridge pouch and the other seventy should be well taken care of and conveyed by the company'.⁹³ Sergeant Roger Lamb of the 23rd Regiment of Foot recalled replenishing his cartridge pouch from the body of a slain member of the Brigade of Guards at the Battle of Guilford Courthouse.⁹⁴

Continental Army soldiers, too, reported a heavy expenditure of ammunition. In 1775 and early 1776, the Continental Army was only able to issue 24 rounds per man.⁹⁵ George Washington wrote to his younger brother, John Augustine, on 31 March that 'I have been here months together with what will scarce be believed – not thirty rounds of musket cartridges of man.'⁹⁶ Here, the commander in chief complains to his younger brother, expressing disbelief at the shortages in the army. Despite this, through October 1776, Washington continued to only call for

---

90   Knox, *An Historical Journal*, vol.1, p.160, vol.2, pp.188–190. 214, 238, 377.
91   Knox, *An Historical Journal*, vol.1, p.314.
92   Thomas Sullivan, *From Redcoat to Rebel: The Thomas Sullivan Journal* (Berwyn Heights: Heritage, 1997), p.102.
93   'Correspondence of General Riedesel,' in *Hessian Documents of the American Revolution*, p.HZ-1974, microfilm.
94   Lamb, *An Original and Authentic Journal*, p.362.
95   LOC: Washington Papers, Major General Philip Schuyler to George Washington, 6 August 1775; General Orders, 17 February 1776; George Washington to Jonathan Trumbull, Sr., 19 February 1776; George Washington to John Hancock, 18–21 February 1776.
96   LOC: Washington Papers, George Washington to John Augustine Washington, 31 March 1776.

issuing men 24 rounds as a result of shortages.[97] Finally, by the second half of the third year of the war, Washington was able to supply his men with more ammunition. Notes from the Philadelphia campaign demonstrate that 40 rounds had become the standard in September and October of 1777.[98] In the same chronological period of the war, the Northern Army under Horatio Gates appears to have possessed slightly less ammunition, calling for between 24 and 30 rounds per man during the Saratoga campaign.[99] Directly before the battles at Saratoga, Horatio Gates also appears to have been requesting buckshot for the Northern Army.[100] The Continental Army under Washington seems to have been able maintain the 40 rounds as standard for the rest of the war.[101] Indeed, by 1781 Continental soldiers began be issued with 50 rounds even if ammunition became a concern in the Southern campaigns after 1778.[102]

From these sources, it is clear that soldiers often used their entire ammunition issue, whether kept in their cartridge pouches or around their person in other ways. One of the few estimates for ammunition usage by an entire army is Mauvillon's calculation of the Prussian Army at in 1742: 'According to my sums, the Prussians fired 650,000 rounds of musketry during their advance at Chotusitz.'[103] When we divide that sum by the roughly 17,000 Prussian infantry at the battle, it seems that the average man fired 38 rounds. With the amount of guesswork, rounding, and estimation involved, caution should be used when using such a sum as evidence in anything but a casual conversation, but it is in the same general area as Johann Jacob Dominicus's reported ammunition usage at the Battle of Paltzig in 1759: 48 rounds.[104]

Eighteenth-century firefights, particularly when conducted at longer ranges, could last for a long time. In both North America and Europe, troops frequently ran short of ammunition and worked feverishly to bring more cartridges into the fight. As armies recognized this issue, soldiers began to carry more and more ammunition and enjoyed close support by ammunition wagons and carts. Firepower increasingly played a dominant role on eighteenth-century battlefields.

---

97   LOC: Washington Papers, General Orders 17 June, 29 June, 30 June, 2 October 1776.
98   LOC: Washington Papers, General Orders, 13 September, 27 September, 5 October 1777.
99   Joseph Johnson, *Traditions and Reminiscences Chiefly of the American Revolution in the South* (Spartanburg, SC: Walker & James, 1972), pp.507, 512, 526–532; *History of the Town of Goshen, Connecticut, With Genealogies and Biographies Based Upon the Records of Deacon Lewis Mills Norton*, p.145.
100 *Horatio Gates Papers*, Microfilm, Reel 5, No. 705, David Mason to General Horatio Gates.
101 LOC: Washington Papers, George Washington to Major General Alexander McDougall, 28 April 1779, General Orders, 16 April 1780, General Orders, 27 April 1780; General Orders (morning orders), 7 June 1780; General Orders, 7 July 1780, General Orders, 13 May 1781.
102 LOC: Washington Papers, Steuben to Jefferson, 5 March 1781; William Heath to George Washington, 23 June 1782.
103 Eleazar Mauvillion, *Histoire de la Dernière Guerre de Boheme* (Amsterdam: Mortier, 1756), pp.100–101.
104 Johann J. Dominicus, *Aus dem Siebenjährige Krieg* (Munich: Beck, 1891), p.62.

Rate of Fire

The final section of this chapter examines how quickly soldiers could fire their muskets across mid-eighteenth-century European armies. It will address the importance of firepower, abilities of troops to fire quickly in a drill square environment, speed of fire in combat, disadvantages of 'quick-fire mania,' and how officers attempted to mitigate those disadvantages. Despite other parade ground or theoretical results, an average of approximately two rounds a minute was quite normal in combat conditions, when soldiers did not engage in 'tap-loading.' This is significant, as it helps explain the length of eighteenth-century battles and the time needed for troops to run low on ammunition.

Oddly enough, Frederick II of Prussia, early in his career seems to have ignored the importance of firepower. In 1748, he encouraged his infantry to attack without firing, and he knew generals would complain 'that I never employ my small arms.'[105] Frederick, it seems, had taken the wrong lessons from the War of Austrian Succession. Shock tactics, not firepower, seemed to be the way forward. It was a much more experienced, seasoned, and defeated Frederick who wrote in 1768:

> The cannon does everything, and the infantry cannot get to grips with cold steel... battles are decided by the superiority of fire. Except in the attack of defended positions, a force of infantry which loads speedily will always get the better of a force which loads more slowly. We have seen this at Rossbach, Liegnitz, and Torgau, and at so many other battles. This is why, I took so much care to introduce speed in loading, after the war, to push soldiers' skill in this area as much as possible. Things are improving, but we must not relax on this point.[106]

After the Seven Years War, then, Frederick had shifted to the belief that firepower and speed of loading decided contests between infantry. Most generals understood this idea, and soldiers were relentlessly taught to fire quickly. Firepower, and sometimes the psychological threat of cold steel, not the cold steel itself, won combats.

First, we need to examine how quickly a soldier could fire on the drill square in peacetime. Across most European nations, the ability to fire four to five rounds seems to have been quite normal. By 1750, most European armies were fascinated with the speed of fire demonstrated by the Prussian troops in the War of Austrian Succession. Austrian military veterans and theorists saw that their troops could fire four to five rounds a minute on the drill square, basically matching the Prussian drill square ideal.[107] Prussian troops were capable of firing six rounds a minute but could not maintain that pace for any length of time. Lossow comments, 'Altogether, it would be too much to expect troops to maintain six rounds a minute for an

---

105 Frederick II, *Military Instruction from the Late King of Prussia to His Generals* (Sherborn: J. Cruttwell, 1818) p.114; Jay Luvaas, *Frederick the Great on the Art of War* (New York: De Capo, 1999), pp.143–144.
106 Frederick II and Volz, *Die Politischen Testamente Friedrichs des Grossen*, pp.146, 148.
107 Duffy, *Instrument of War*, p.441; Duffy, *The Army of Frederick the Great* (2020), pp.146–147.

extended time.'[108] The British achieved a similar level of drill-square proficiency in this regard, with veteran troops being able to fire four rounds a minute.[109] The Russians, too, used Prussian emphasis speed of fire when training their troops in the 1750s.[110] Some sources thought this too ambitious. Prussian *Generalleutnant* Ludwig von Lossow argued that, 'in ... European armies only the most practiced parts of them can fire four times in a minute's time. Usually, only three shots come are fired: watch and see!'[111] Regardless, by the 1780s the cylindrical ramrod and the conical touchhole (technological assists to loading) had increased the drill-square rate of fire to six rounds fired with a seventh round loaded.

However, as with almost all aspects of military life, there was a severe disparity between what soldiers could accomplish on the peacetime drill square and the battlefield. It seems that troops in combat fired more slowly. Two rounds a minute is a very believable figure for well-trained, veteran troops in combat. Austrian army officer Jakob Cogniazzo gives us a window into this idea:

> Now, how many rounds of rapid fire do you think he can loose off in a minute when he is in a minute when he is in this condition? At least five a minute? That is certainly the norm for fire on the drill square, which conjures up visions of enemy corpses by the thousand. But, when we consider all the encumbering burden of the soldier... taking everything into due account, it would be optimistic to suppose that he fires as many as one or at the most two rounds a minute [in combat].[112]

One of the differences between drill square 'minute firing' and real combat was its duration. The real test was how long soldiers could keep firing at a high rate. It seems British troops could fire two to three rounds a minute for a sustained amount of time.[113] Likewise, it seems that during combat the Prussians could fire three rounds per minute, and they could keep up this pace for extended periods of time.[114] The Russians loaded slower, perhaps one to two shots a minute, as a result of the additional time it took to load their buckshot rounds. An observer during the Zorndorf campaign recalled:

> …every Russian infantryman loads a musket ball and an equal pack of buckshot. There are between 7–9 of these in a linen packet in the form of grapeshot. As a result of this, the Russians load quite slowly, as a Jäger

---

108 Lossow, *Denkwürdigkeiten zur Charakteristik der preussischen Armee*, p.265.
109 Richard Holmes, *Redcoat: The British Soldier in the Age of Horse and Musket* (New York: W.W. Norton, 2002), p.199.
110 Christopher Duffy, *Russia's Military Way to the West* (London: Routledge & Keegan Paul, 1980), p.62.
111 Lossow, *Denkwürdigkeiten zur Charakteristik der preussischen Armee*, p.264.
112 Jakob de Cogniazzo, *Freymüthiger Beytrag zur Geschichte des Österreichischen Militairdienstes* (Frankfurt and Leipzig: 1780), p.147.
113 John Houlding, *Fit for Service: The Training of the British Army, 1715–1795* (Oxford: Claredon Press, 1981), p.194.
114 Duffy, *The Army of Frederick the Great* (2020), pp.146–147.

loads his rifle. In the time it takes the Russians to load their weapons, the Prussians have fired three times.[115]

With this in mind, it seems that troops carrying 30 rounds would run low on ammunition after 10–20 minutes, while troops carrying 60 rounds of ammunition would run low after 30–45 minutes. Ulrich Bräker, at the Battle of Lobositz, noted that after firing off 60 rounds in succession, his musket became so hot that he had to carry it by the sling.[116]

More important than the actual number of shots was the ability to confer a comparative advantage to troops who could load and fire more quickly than their opponents. An Austrian officer noted that the Prussians had 'the important and extraordinarily, significant advantage of being able to get off three rounds to every one of the Austrian infantry. This is conceded by all impartial and well-informed men who have seen it with their own eyes.'[117] Prussian officer Ernst von Barsewisch recalled of the Battle of Liegnitz: 'Now I commanded, 'platoon: ready! present! fire!' Then the remaining part of the battalion followed, whereupon we blasted away for a length of time. The enemy, however, did not withdraw, but also fired vigorously. But we loaded more speedily and had devastated the enemy with our first volley.'[118]

Instilling troops with the need to fire quickly at the expense of all other factors had a number of downsides. Again, *Generalleutnant* von Lossow comments that officers 'forget that the musket barrel becomes too hot to hold after two minutes... the soldiers inevitably acquired bad habits when they were put through the rapid-fire drill every day, like failing to ram the loads firmly home, neglecting to aim, raising the barrel too rapidly after they had loosed off, and so on.'[119] It seems there might have been a downside to this 'quick-fire mania.'

In addition to these bad qualities, soldiers might begin to 'cheat' in other ways. Of these, the most infamous was the so-called 'tap-loading.' In 1726, General Hawley complained that 'the German and Dutch foot might be brought to ram their cartridge every time on service, for ought I know, but by the nature of our men I believe it impossible to bring them to it.'[120] The Austrians engaged in tap-loading at Mollwitz in 1741, where it robbed their musket discharges of lethal force. The French used tap-loading at Lauffeld in 1747, and the British engaged in the practice at Hubbardton in 1777.[121]

---

115 Anon., *Besondere Merkwürdigkeiten und Anekdoten aus Neudam in der Neumark* (Salzwedel: J.C. Schuster, 1758), p.29.
116 Ulrich Bräker and Johann Heinrich Füssli, *Lebensgeschichte und natürliche Ebentheuer des Armen Mannes im Tockenburg* (Zürich: Orell, Gessner, Füssli,1789), p.150.
117 Report of Lt. Colonel Rebain, 10 May 1758, Vienna Kriegsarchiv CA 1758 III, p.1.
118 Barsewisch, *Meine Kriegs-Erlebnisse*, p.113.
119 Lossow, *Denkwürdigkeiten zur Charakteristik der preussischen Armee*, p.266.
120 Quoted in G. Tylden, 'Notes on Musket and Rifle,' in *Journal of the Society for Army Historical Research*, vol.31, no.136 (1953), pp.88–89.
121 Anon., *Denckwüdiges Leben und Thaten Beruehmeten Herren Johaan Daniels von Menzel*, p. 80; Blackmore, *Destructive and Formidable*, p.104.; Anbury, *Travels through the Interior Parts of America*, p.333.

Military authorities from various nations were aware of these problems and attempted to ensure that soldiers fired with speed and accuracy on the day of battle. Soldiers in almost all European nations practiced firing at targets to ensure accuracy. Troops in Central European armies were carefully taught to aim their muskets with reference to their distance from the target.[122] British general James Wolfe argued, 'there is no necessity for firing very fast; a cool and well-leveled fire, with the pieces, carefully loaded, is much more destructive and formidable than the quickest fire in confusion.'[123] Despite this, numerous military theorists believed that though disadvantageous, there were times when quick-fire and tap-loading outweighed these concerns.[124] Officers and military theorists attempted to carefully weigh and balance speed of fire with accuracy.

So, what can we assert regarding the rate of fire among regular troops in the mid-eighteenth century? These troops could fire at prodigious rates on the drill square but often fell short of this ideal on the battlefield. Sometimes, this led to the disadvantages associated with 'quick-fire mania' or the danger of tap-loading. As a result, officers attempted to find a happy medium: soldiers who would fire at speed but retain accuracy.

Conclusion

On eighteenth-century battlefields from Parma to Pomerania, and Lobositz to Long Island, soldiers sometimes fought in ways that ignored the drill manuals and carefully crafted advise treatises of their officers. Instead, these men obeyed their officers where possible, but often took battles into their own hands, firing without orders at greater ranges than their officers preferred for extended duration, using up large quantities of ammunition at a relatively high rate of speed. These troops endured the psychological hardship of holding fire where possible and were begged, cajoled, and threatened into holding their fire as long as they could bear.

In some ways, these officers were undoubtedly correct in their opinions. A supremely disciplined force could manoeuvre effectively on the battlefield. In Frederick the Great's view, constantly disciplining the soldiers to march well enabled the line to advance 'without floating and breaking.'[125] With perfectly aligned troops, it was vastly easier to wheel, advance in order, and maintain fire control. By maintaining order in the ranks, officers hoped that they could use the weapons and tactics of the time to their maximum potential. Despite this, a wide gulf remained between what officers hoped was theoretically possible and what soldiers actually practiced on the battlefield.

What is clear from the sources is that Johann Gottfried von Hoyer correctly described the situation at the end of the century. In his view, it was impossible to construct an 'robotic shooting-machine' even among the vaunted Prussian troops.

---

122 Duffy, *Instrument of War*, p.442; Duffy, *The Army of Frederick the Great* (2020), pp.147–148.
123 James Wolfe, *General Wolfe's Instructions to Young Officers* (London: J. Millan, 1780), p.49.
124 Blackmore, *Destructive and Formidable*, pp.104–105
125 Frederick II and Volz, *Die Politischen Testamente*, p.148.

Instead, the Prussian skill in loading the musket provided advantages in the firefight, as they loaded comparatively faster than their Austrian and Russian opponents. This speed and dexterity, rather than robotic obedience or an ability to consistently platoon fire, gave the Prussian troops the small comparative advantages they actually enjoyed over their opponents. Having turned to one way that soldiers attempted to shape battle, we will now turn to their efforts to survive the deadly world of eighteenth-century battle.

8

Surviving Combat: Taking Cover, Running on the Battlefield, and Melee

> *Of our own accord we ran up to the cover of the young pine grove, but as soon as we ran up out of it, we were greeted by the enemy's artillery fire.*[1]
>   Marcin Matuszewicz, Brzeg Infantry Regiment, 1734

Few eighteenth-century battles combine the themes of this chapter better than the short skirmish at Clifton on 18 December 1745. Fought in difficult terrain, it involved troops moving at high speed and, despite the best efforts of the retreating party, a brief melee. After the skirmish, an anonymous Jacobite officer sat to record his memoirs of the skirmish. In his view the Jacobites had been initially deployed to cover themselves against fire, as 'our General Lord George Murray were planted as follows: Glengaries at the back of a stone dyke on the right, Stuart of Appin and mine on the left at a thorn hedge.'[2] Despite this relatively advantageous position, Murray, as so many generals have, saw a position just a little bit better: 'the general spying another hedge at more than a gunshot distance and as it was much nearer the enemy than where we were which he thought more commodious ordered the Stuarts and us to quit our first stations and endeavour to get ourselves possessed of that hedge before the enemy could come up.'[3] The stage was set for a memorable encounter between two parties of infantry in the hedges around Clifton.

Murray's troops had no sooner left their covered positions when a party of Government dragoons, 'as if entrenched to the teeth by means of a deep ditch and toss hedge fired most warmly and briskly upon us.'[4] Somewhat embarrassed, the Jacobite forces noted that they 'were going on with a pretty swift pace and without any cover.'[5] Caught in the open without cover, the Jacobite troops found that they had few options to but to return fire. Then:

---

1 Marcin Matuszewicz and Adolf Pawiński, *Pamiętniki Marcina Matuszewicza: Kasztelana Brzeskiego-Litewskiego, 1714–1765* (Warsaw: Gebethner and Wolffa, 1876), vol.1, p.52.
2 RA: Cumberland Papers Main Series, 8/190, Account of the Action at Clifton, 18 December 1745.
3 RA: Cumberland Papers Main Series, 8/190, Account of the Action at Clifton, 18 December 1745.
4 RA: Cumberland Papers Main Series, 8/190, Account of the Action at Clifton, 18 December 1745.
5 RA: Cumberland Papers Main Series, 8/190, Account of the Action at Clifton, 18 December 1745.

...in the moment we loaded our pieces made a shout, drew our swords, and rushed over the ditch and hedge and upon them... they endeavoured to fly with precipitation to their main body were broke a great many of our swords on their scullcapes [metal inserts or helmets in the dragoon's hats] each of them being armed with fusils and followed them within pistol shot of the grand army. We were then ordered by the general to retire and charge again to be ready against a second attack in case any more should venture out against us.[6]

The evening skirmish at Clifton then demonstrates a few key features of eighteenth-century infantry warfare. First, both sides desired to utilize cover if tactically acceptable. Stone dikes, ditches, and hedges were all valuable assets in a firefight. If troops were in the open, particularly if being fired upon, they were not afraid to move out of that situation at a swift pace. Desperate situations, such as being fired on in the open by enemies in cover, could cause a swift moving attack designed to chase the enemy away from the battlefield. If the enemy was defending an entrenched position, a melee might result, but more frequently, one side would flee from close combat.

The action at Clifton shows an eighteenth-century combat that was dynamic and driven by a search for superior cover and firing positions. This flies in the face of the stereotypical image of eighteenth-century battle where the popular imagination saw British officers doffed their hats and said to their chivalrous opponents supposedly saying, 'gentlemen of France, fire first!' According to this view, soldiers move to the beat of drums with perfect clockwork precision, and could not imagine combat occurring outside of a geometric, mechanical framework. As a result of the principal weapons system of the time, the smoothbore musket, soldiers usually fought in linear formations. In Europe, these linear formations were usually in close order. However, as other chapters of this book have suggested, there was a difference in the eighteenth century, as there is today, between the parade ground and the battlefield. These officers and soldiers were not unthinking automata who were incapable of performing any tasks but those learned on the drill square. Rather, under the direction of junior officers, they frequently modified their actions to fit with local conditions.

Taking Cover

Soldiers facing deadly incoming fire frequently tried to utilize cover in eighteenth-century battles. First of all, from the available sources, it seems as though this practice was incredibly common. So common, that it might even be considered a normal (if not universal) practice when on the defensive in the 1740–1783 era. Oddly enough, unlike the shallow, open-order formations used by the British, lying down under fire does not seem to have developed in North America. Instead, this practice was common on European battlefields from the 1730s, and as noted in the introduction,

---

6   RA: Cumberland Papers Main Series, 8/190, Account of the Action at Clifton, 18 December 1745.

was also practiced in the seventeenth century. Also, it should be carefully noted: by and large, the troops performing this tactic are battalion company soldiers, not light infantry of any sort. As usual throughout this work, where possible sources from light infantry or provincial units have been avoided, in order to establish that infantry of the line employed this tactic. Austrian, British, French, and American soldiers – sometimes under orders, sometimes of their own accord – stooped low or lay down in order to present their enemies with a smaller target.

The first account of this type of behaviour comes, as we have seen, from the War of Polish Succession at Parma at Guastalla, but this is not the only incidence of troops using cover in the 1730s. During an assault on the town of Ochakiv in 1737, Russian troops fought with their Ottoman opponents in the gardens and suburbs of the town, with both sides utilizing the built-up area to fire on the enemy from cover. The Ottoman troops had essentially, in the words of Russian officer Christoph Hermann von Mansteinm, 'two gardens converted to redoubts.'[7] After driving the Ottomans from these positions, the Russian army utilized them to fire on the town. Field Marshall von Münnich ordered the Russian troops to leave the protection of the built-up gardens and 'come from behind the redoubts, and fire without cover.'[8] The direct commander lodged a protest at this order, but the troops obeyed for a time, before disaster struck: 'At length, the troops growing disheartened... retired in great confusion into the gardens and redoubts which they had occupied the previous night.'[9]

British troops appear to have employed this tactic in almost every major battle of the War of Austrian Succession. Lieutenant Colonel Russell of the Guards describes the British and French infantry stooping low at the Battle of Dettingen in 1743, and in his words: 'our foot almost kneeled down by whole ranks, and so fired upon 'em a constant running fire.'[10] Russell describes a pattern by which both sides would wait until individual enemy soldiers would rise from a stooped position to load or fire, and then pick off the exposed individual enemy soldiers.[11] Although Russell's description of events is suspect because he took no significant part in the fighting himself, when weighted together with the other evidence from the War of Austrian Succession, it becomes convincing.

A young James Wolfe, describing the French at Dettingen, recalled that as the British prepared to give fire: 'As soon as the French saw we presented they all fell down, and when we had fired they all got up, and marched close to us in tolerable good order, and gave us a brisk fire, which put us into some disorder and made us give way a little.'[12] This clearly shows a practice of alternating cover (when the enemy was firing) and advancing and firing (when the enemy was reloading). The result

---

7 Christoph Hermann von Manstein, *Memoirs of Russia from the Year 1727 to the Year 1744* (London, Becket and Hondt, 1773), p.151.
8 Manstein, *Memoirs of Russia*, p.153.
9 Manstein, *Memoirs of Russia*, p.154.
10 Historical Manuscripts Commission, *Report on the Manuscripts of Mrs. Frankland-Russell-Astley*, p.260.
11 Historical Manuscripts Commission, *Report on the Manuscripts of Mrs. Frankland-Russell-Astley*, p.278.
12 Wilson, *The Life and Letters of James Wolfe*, p.37.

was relatively small losses on the part of a unit, even while that unit was advancing to the attack against a firing enemy. Wolfe recalled wistfully that after the French had thrown themselves to the ground, 'We did very little execution.'[13]

The next major French-British engagement of the War of Austrian Succession, the Battle of Fontenoy, contains a number of instances of British soldiers employing cover to avoid enemy fire. The first instance is that of the Sempill's Highlanders, more famously later known as the Royal Highland Regiment. Under the orders of Colonel Robert Munro, the Highlanders were permitted to engage 'in their own way of fighting.'[14] When the French would prepare to fire a volley, the Highland troops would 'clap to the ground,' allowing the French fire to pass over them, and 'instantly as soon as it was discharged, they sprung up, and coming close to the enemy, poured in their shot.'[15] The fact that sources describe this as being 'their own way of fighting' might seem to imply it is a Scottish peculiarity, but this is not totally accurate.

Sampson Staniforth, a private soldier (probably in Bligh's Regiment, which would later be known as the 20th of Foot) described his experience of combat at Fontenoy as follows:

> We marched up boldly; but when we came close to the town of Fontenoy, we observed a large battery ready to be opened on us. And the cannon were loaded with small bullets, nails, and pieces of old iron. We had orders to lie down on the ground; but for all that, many were wounded, and some killed. Presently after the discharge we rose up, and -marched to the first trench, still keeping up our fire.[16]

So, like the Highlanders, regular British infantry would lay down under the threat of severe enemy fire. Later in the war, at the Battle of Rocoux, Staniforth again gives us a window into the defensive nature of laying down under fire. He begins, 'We English posted ourselves in some gardens and orchards, which were some little cover,' and other soldiers who were there note that the British tried to fortify the hedgerows by filling them with dirt.[17] As the French forces marched on Staniforth's position:

> Here we lay, waiting for orders to retreat to our army... While we lay on our arms, I had both time and opportunity to reprove the wicked. [Staniforth was a deeply religious man] By this time the French came very near us, and a cannon-ball came straight up our rank. But, as we were lying upon the ground, it went over our heads. We then had orders to stand up and fire.[18]

---

13 Wilson, *The Life and Letters of James Wolfe*, p.37
14 Philip Doddridge, *Some Remarkable Passages in the Life of the Honourable Colonel James Gardiner* (London: 1794), p.267.
15 Doddridge, *Some Remarkable Passages*, p.267.
16 Thomas Jackson, *The Lives of Early Methodist Preachers Chiefly Written by Themselves* (London: Wesleyan Conference Office, 1872), vol.4, p.172.
17 Jackson, *The Lives of Early Methodist Preachers*, vol.4, p.135.
18 Jackson, *The Lives of Early Methodist Preachers*, vol.4, p.136.

So, in at least three major battles of the War of Austrian Succession, British troops presented their enemies with smaller targets to avoid enemy firepower. The French also employed the use of cover at the Battle of Lauffeld on 2 July 1747. Allied cavalry received 'a sharp fire from the foot, which [the French] had posted in a hollow way, and some hedges.'[19] They were matched by the Austrians in the Central European theatre of conflict during this war. In a skirmish near Landeshut on 21 May 1745, Prussian commanders reported that 'the enemy infantry and grenzers fired from cover.'[20] The Austrians seemed to have learned one major lesson of the fighting at Parma and Guastalla: built-up terrain could provide a tactical advantage. Thus, an officer of the Margrave Carl Regiment reported of the Battle of Hohenfriedberg: 'But as soon as our artillery, which like our other regiments, consisted of two twelve-pounders, started to damage the enemy, they retreated. After firing a few volleys of small arms, they fell back behind farmstead walls, and defended themselves there for a little while longer.'[21]

In the War of Austrian Succession and Seven Years War, Prussian units were not afraid to seek cover in villages. In the War of Austrian Succession, Prussian dragoons barricaded themselves in houses and fired from windows.[22] In Piedmont, French and Austro-Piedmontese troops skirmished in September of 1747. Here, the lessons of Parma and Guastalla remained in the forefront, with Austrian officers noting village terrain such 'the village of Arche, where there is a church and walls proper for a breastwork.'[23]

This trend continued during the Seven Years War. During the raid on Cherbourg in 1758, Corporal Todd of the 12th Regiment of Foot recalled lying down in order to prevent friendly-fire incidents: 'And as soon as we got ashore, we lay down close upon the Beach, near the water edge, that our ships might fire over us in case the Enemy Advance to make any Attack upon us'.[24] British soldiers laid down, not only to make themselves smaller targets but also to facilitate supporting fire in certain contexts.

During the Seven Years War, Prussian units were not afraid to seek cover in villages. In September of 1761, regular infantry under the command of *Generalleutnant* Dubislav Friedrich von Platen stormed the Russian encampment near Gostyn, and the Görne Grenadier Battalion 'fired from windows' of the Kloster Meister.[25] This was confirmed in the reports of multiple officers.[26] Likewise, Prussian officers fighting in Pomerania noted that their Swedish opponents found success because they fought 'defensively behind [garden] walls.'[27] At a skirmish in 1762, a

---

19 RA: Cumberland Papers, Main Series 24/22d, Relation of the Action at the Village of Val.
20 Anon., *Sammlung ungedruckter Nachrichten so die Geschichte der Feldzüge der Preußen von 1740 bis 1779* (Dresden: Walther, 1785), vol.4, p.118.
21 Anon., *Sammlung ungedruckter Nachrichten*, vol.1, p.332.
22 Anon., *Sammlung ungedruckter Nachrichten*, vol.4, p.217.
23 TNA: SP 87/23, f.382, Lt General Wentworth to the Duke of Newcastle, Demont[e], 22 September 1747.
24 Todd, *The Journal of Corporal Todd*, p.168.
25 Anon., *Sammlung ungedruckter Nachrichten*, vol.3, p.217.
26 Anon., *Sammlung ungedruckter Nachrichten*, vol.3, p.8.
27 Anon., *Sammlung ungedruckter Nachrichten*, vol.5, p.217.

Prussian dragoon officer noted that the enemy infantry, 'particularly the Guilay and Esterházy Regiments... made a heavy fire from the walls and the windows' of a hospital.[28] Another Prussian officer noted the willingness of Austrian troops to fire with small arms, and even set up batteries, in villages and suburbs using garden walls.[29] During the 1759 campaign in Franconia, Austrian and Reichsarmee troops defended the suburbs of Hof, and the Cronegk Regiment tried to hold off enemy troops from 'a wall at the edge of a wood.'[30] Facing a battery of Russian cannon at Paltzig, a platoon of the Prussian Anhalt-Dessau Regiment took cover behind trees and fired at the enemy from there.[31]

In July 1758, during the Swedish war against the Prussians in Pomerania, Swedish *Kapten* Vilhelm Armfeldt of the Dalarna Infantry Regiment recalled being bombarded by Prussian artillery:

> While we were there and saw how the farm nearby was burning brightly, two [Prussian] soldiers came from the redoubt, sneaking through the grain to some houses behind us. The [guard] who stood in a small garden, saw them and let us know... stating that they were hussars, since he only saw their caps from time to time as they walked through the grain. I let some men from the [guard] advance...but they [the Swedish troops] barely arrived at the farm closest to the redoubt before they began shooting at the two fellows, which saw us and ran back as fast as they could. When the redoubt heard our shots and saw where they came from, they started shooting in our direction, and one bomb was so well aimed that it struck barely [16 feet] from us, exploding. The men threw themselves on the ground, since they were closest, but we, Captain Hertz and I, remained standing. The bomb threw soil and burning debris at us, but no one, praise the Lord–was hurt.[32]

Here, Armfeldt clearly demonstrates the tension between choosing to take cover and not: two officers seem to believe that might be dishonourable to take cover, while the enlisted men who are closer to the explosion throw themselves on the ground. Officers and men who were anxious to display their courage in the face of the enemy seem to have not taken cover even in deadly situations like Armfeldt found himself in above.

In North America, British soldiers frequently laid down under fire. John Knox frequently describes this practice during the 1759 campaign at Quebec. He describes laying down in the course of manoeuvring against the enemy, during

---

28 Anon., *Sammlung ungedruckter Nachrichten*, vol.5, p.398.
29 Anon., *Sammlung ungedruckter Nachrichten*, vol.1, p.565.
30 Isaac Jakob Petri, Alister Sharman and Neil Cogswell, 'A Comprehensive Journal Relation of the Campaign of His Royal Highness Prince Henry against the Combined Austrian and Empire Army', Unpublished Archival Report. (2000), pp.2, 19.
31 Jakob Friedrich von Lemcke and R. Walz (ed.), 'Kriegs-und Friedensbilder aus den Jahren 1754–1759', *Preußische Jahrbücher*, vol.138 (1909), p.36.
32 Krigsarkivet (KrA): 0391/0/33, Entry for 21 July 1758, Handjournal of Major Vilhelm Armfeldt.

siege operations when taking fire from enemy artillery and on the battlefield against enemy infantry.[33] In addition, the British infantry used this tactic at the Battle of La Belle Familie, as suggested at the beginning of this article.[34]

Unsurprisingly, the practice continued during the American War of Independence. Again, in this era, it seems to have been a response to coming under artillery fire. At the Battle of Harlem Heights, multiple British diarists record that they lay on their arms after coming under fire by American artillery batteries. Thomas Sullivan, with the 49th Regiment of Foot, recalled, 'The Cannonading continued at both sides for an hour... All that time out Brigade i.e. 2d., lay upon our Arms in a field of Indian corn...' Sullivan describes this practice again at Brandywine in 1777.[35] Also in 1777, Ensign Thomas Glyn of the Brigade of Guards reported, 'the Enemy advanced with two pieces of Cannon & began to cannonade us, when we were ordered to lay down and being covered by the ground, no loss ensued...'[36] Once again, even while employing the quick aggressive tactics outlined by Matthew Spring, the British were not afraid to take cover if the situation demanded it.

As in the Seven Years War, this British practice extended far beyond North America. At the Battle of La Vigie on St Lucia in 1778, British soldiers repeatedly took cover to avoid French firepower. Major George Harris of the 5th Regiment of Foot recalled, 'My gallant friend, now no more, Captain Shawe of the 4th. Company, was ordered by me to make his men lie down, and cover themselves with brushwood as much as possible, to prevent them being seen as marks.'[37] Later in the war in the Caribbean, Lieutenant the Honourable Colin Lindsay of the 55th Regiment also reported that his soldiers took cover in the course of the fighting. Thus, across most times and places, infantry units of most armies were willing to take cover from enemy fire in a variety of ways. The British utilized this practice most often by laying down in the open, but essentially all eighteenth-century infantrymen could fight from windows, walls, and doors in a built-up environment.

Running on the Battlefield

Soldiers often moved quickly on the battlefield as the situation demanded, often moving at a run in moments of crisis. To students of the British Army in the American War of

---

33 John Knox, *An historical journal of the campaigns in North America for the years 1757, 1758, 1759, and 1760* (London, 1769), vol.1, pp.302, 308, 321; vol.2, p.70.
34 TNA: PRO 30/8/49, Eyre Massey on the Battle of La Belle Familie, p.93. The best modern description of this battle is in: Stephen Brumwell, *Redcoats: The British Soldier and War in the Americas, 1755–1763* (Cambridge: Cambridge University Press, 2002), p.252.
35 Thomas Sullivan, *From Redcoat to Rebel: The Thomas Sullivan Journal* (Berwyn Heights: Heritage, 1997), pp.67, 131.
36 Ensign Thomas Glyn, 'Ensign Glyn's Journal on the American Service with the Detachment of 1,000 Men of the Guards commanded by Brigadier General Mathew in 1776', *Varnum Lansing Collins, Revolutionary War papers, 1913–1932* (Princeton: Manuscript Department, Princeton University Library), p.30.
37 Lord Lindsay, 'Narrative of the Occupation and Defence of the Island of St. Lucie against the French, 1779', in Lord Lindsay, *Lives of the Lindsays, Or A Memoir of the Houses of Crawford and Balcarres* (London: John Murray, 1849), pp.344–346.

Independence, this should not come as surprising news. Matthew H. Spring has recently shown that British soldiers fought unconventionally in North America. However, this trend goes beyond North America and the American War of Independence. Eighteenth-century soldiers were often rational actors and made competent decisions based on the needs of the moment and battlefield crises. This willingness to respond quickly to developing factors was not a feature of the British or Prussian armies, but common to almost all eighteenth-century militaries. To understand this type of behaviour, it is very important not to confuse the speedy flight of troops on the battlefield with men intentionally moving at speed in a more controlled way.

As the first chapter discussed, Jacques Mercoyrol de Beaulieu, in conversations with French veterans of the War of Polish Succession in the 1730s, recalled the types of battlefield movements possible in heat of the moment: 'In these battles, (which were all victories) the Picardie Brigade... They went almost at as in a race, the quickest arrived first; this eagerness awakened courage and gave new strength to the tired fighters.'[38] Moving quickly, perhaps even at the run, was common enough on European battlefield when speed was required.

The Prussian army was famous for its attention to drill and parade excellence. It may be somewhat surprising, therefore, to learn that Prussian soldiers under Frederick the Great occasionally moved with great speed. The official war journal of the Prussian Fusilier Regiment of Jung-Braunschweig makes this clear. In describing the Battle of Prague in 1757, the journal describes both the Prussian regiment and their Austrian opponents moving at speed, two instances of quick movement: 'In order to reach the position of advance, we had to pass a long dam, which delayed us. So, in order arrive at the correct time, we had to run past the village of Arhem: the regiment was not in perfect order. The enemy [Austrians] were already advancing on us at the quick step, and we engaged them.'[39]

Johann Jakob Dominicus, a musketeer in Frederick II's army, also remembers running at Prague. He wrote, 'our left wing had its work cut out for it, and we had to run with energy, in order to get under the enemy guns.'[40] In Prussia, speed and reaching the appointed position at the right time (so as not to leave a gap in the battle line) took precedence over moving in step and keeping good order. Prussian veteran Georg von Berenhorst asserted that commanders who lost time by correcting minor irregularities in dress were punished, as maintaining a healthy and cohesive battle line was more important than keeping in parade appearances.[41]

Generalleutnant Ludwig Matthias von Lossow recalled that as soon as the firing began, 'the orderly lines fell by the wayside, as they did whenever troops advanced through terrain, as it is rarely open enough to permit lines of this size. Units only needed to keep in contact with the other parts of the line, that was the main thing, as

---

38　Jacques de Mercoyrol de Beaulieu, *Campagnes de Jacques de Mercoyrol de Beaulieu, capitaine au régiment de Picardie: 1743–1763* (Paris: Libairie Renouard, 1915), p.184
39　Anon., *Sammlung ungedruckter Nachrichten so die Geschichte der Feldzüge der Preußen von 1740 bis 1779* (Dresden: Walther, 1785), vol.4, p.118.
40　Johann J. Dominicus, *Aus dem Siebenjährige Krieg* (Munich: C.B. Beck Verlag, 1891), p.16.
41　Georg Heinrich von Berenhorst, *Betrachtungen über die Kriegskunst* (Leipzig: Fleischer, 1798), vol.1, p.221.

the experienced army of Frederick II knew well.[42] Keeping in contact with other units and keeping up the advance, it seems, were higher priorities than keeping in perfect close order. A Dutch officer in Prussian service recalled a nighttime skirmish in 1757:

> Because of the darkness of night, we had difficulty distinguishing our troops from the enemy, so Lt. York received orders to reconnoitre the enemy with two platoons. They fired on him. Captain Rodig, who had been fired upon, rode out to these flanquers... His pickets behaved bravely, and kept up an orderly fire in the manner of platoons. They then rushed forward on the command, 'Marsch! Marsch!' The enemy took flight with haste.[43]

David Dundas, one of the principal English-language observers of the Prussian army in the later part of Frederick II's reign, recalled this after seeing the Prussian manoeuvres in the 1780s:

> Quick movements, which considerable columns or lines of infantry [Frederick II], considers as impracticable and ruinous from the hurry and disorder that must thence ensue... but brigades, or smaller divisions of the line [such as regiments or battalions], occasionally lengthen their step, and move on with rapidity at the moment of attack.[44]

In delegating more responsibility to junior officers, the British facilitated quick movement even more in North America during the American War of Independence. Matthew H. Spring exhaustively shows that British troops moved at a kind of jog or trot:

> The King's troops, 'briskly marched up to' the enemy at Long Island, 'briskly ascended' Chatterton's Hill, 'advanced fearlessly and very quickly' at Brandywine, came on at Bemis Heights at a, 'quick step' stormed for Clinton, 'with as much velocity as the ground would admit,' and 'after a very quick march moved up briskly' against the enemy at Monmouth. Likewise, in the South, the redcoats, 'marched forwards briskly, or rather rushed with great shouts,' at Savannah, were observed 'advancing rapidly' at Briar Creek, and 'rushed on with the greatest rapidity' (or 'as fast as the ploughed fields they had to cross would admit') at Spencer's Ordinary. Most expressively of all, one rebel militiaman at the battle of Cowpens later recalled, 'the British line advanced at a sort of trot with a loud hallo. It was the most beautiful line I ever saw,' while another reported that the King's troops, 'advanced rapidly as if certain of victory.'[45]

---

42 Ludwig Matthias von Lossow, *Denkwürdigkeiten zur Charakteristik der preussischen Armee, unter dem grossen König Friedrich dem Zweiten. Aus dem Nachlasse eines alten preussischen Offiziers* (Glogau: Carl Heymann, 1826), p.242.
43 Anon., *Schreiben des Holländischen Volontairs* (Dresden: Publisher Unknown,1757), vol.9, p.5.
44 David Dundas, *Principles of Military Movements Chiefly Applied to Infantry* (London: Cadell, 1788), p.9.
45 Spring, *With Zeal and With Bayonets Only*, p.146.

All of the quotes in the above paragraph come from observers present at the battles, and Spring provides a detailed footnote for those looking to track them down. Thus, in the American War of Independence, British soldiers moved quickly as part of usual practice, rather than speeding up when the circumstances demanded it. Roger Lamb recalls that the British moved forward at Guilford Courthouse 'in excellent order, at a smart run, with arms charged.'[46] At the same battle, the normally slower Hessians in the von Bose Regiment joined the British advance with speed:

> After quickly laying aside our tornisters and everything that could impede a soldier, the 71st and von Bose received orders to move forward and attack the enemy... We had not advanced more than 300 yards when we found a deep ditch in front of us, with tall banks and full of water. After crossing it with difficulty, we then came to a fenced wheat field; on the other side of this field 1,500 continentals and militia were deployed in line... I formed the battalion into line with the greatest of speed and we ran to meet the enemy in tolerable order.[47]

Other German allies of the British, the Brunswickers under Baron Riedesel, appear to have moved at speed during the culmination of a flank attack during the Battle of Freeman's Farm on 19 September 1777. Hearing the British troops engaged with the rebel Americans, Riedesel moved out 'as quickly as possible' and launched his final attack 'at the quick-step.'[48]

It is important that we do not overdraw these examples. In the eighteenth century, most European commanders valued ordered bodies of men, and preferred to attack in an orderly fashion. However, in moments of crisis, European junior officers frequently took it upon themselves to move bodies of men at speed in order to contain crises or take advantage of conditions. The British took a decidedly different approach in the American War of Independence. In that conflict speed was institutionalized in the British Army. Whether in Europe or North America, these soldiers were not automata; they moved at the speed demanded by the situation.

Avoiding Melee

It seems that mid-eighteenth-century soldiers frequently fired their entire ammunition load, engaged in firefights at a much greater distance than is usually assumed, and sometimes fired at will rather than fire by platoon or division. What does all this mean for eighteenth-century combat? If troops fired quickly, it could help to explain the ammunition usage, and the range of combat can help historians make sense of

---

46   Lamb, *An Original and Authentic Journal of Occurrences during the Late American War*, p.361.
47   HSM: Best. 4h Nr. 3101, Johann Christian Du Buy, Raports vom Oberst Lieut. du Buy Regts v. Bose zu der General Lieutenant v. Knyphausen, p.117.
48   Max von Eelking, *Leben und Wirken des Herzolich Braunschweigischen Generals-Lieutenants Friedrich Adolph Riedesel* (Leipzig: Wigand, 1856), vol.2, pp.149–150.

the large numbers of rounds expended versus small number of casualties. But what does all this mean for melee (hand to hand or close) combat? Did eighteenth-century soldiers engage in melee combat frequently? If the Swedish army during the Great Northern War and the British Army during the American War of Independence preferred bayonet attacks, surely there was a good deal of hand-to-hand combat?

Melee combat occurred, but it was perhaps less frequent than might be initially imagined. There are many famous examples: Culloden, Bunker Hill, and Guilford Courthouse to name a few. When fighting against the Ottoman Empire, both Russian and Austrian soldiers reported fierce hand-to-hand struggles. Many military commanders, at some point in their careers, seemed to prefer an *armes blanche* or cold-steel attack. Frederick the Great advocated for this idea in the early Seven Years War. Alexander Vasilyevich Suvorov famously stated that 'the bullet is a mad thing, put your trust in the bayonet.' With these ideas spreading, surely hand to hand combat was frequent?

In reality, hand-to-hand, or melee combat, was often limited to a select number of places on the battlefield. When enemy troops appeared to make a serious advance into close range with bayonets, defending troops often melted away. That is why the Swedish Caroliner and British redcoats proved so effective on their respective battlefields. It also helps to explain why, when things went wrong for troops making a charge with cold steel, they went very wrong (such as at Poltava in 1709 and Cowpens in 1781). Troops did experience hand-to-hand combat, but firepower (at range) was the order of the day. In William Dalrymple's 1782 essay on tactics, in the case of infantry, he asserted:

> There is probably not an instance of modern troops being engaged in close combat... the bayonet can be of little utility by way of impulsion in the field... these defects in modern infantry prove the impracticability of two battalions, opposed to each other, being brough in the open field to close encounter: one body must give way before they get into action.[49]

Though Dalrymple is exaggerating for effect, we would be wise to take his point. In the open field, when flight was a possibility, it was rare for two battalions of infantry to cross bayonets. Nonetheless, other types of melee combat did occur and are worth discussing. Even in nations where military theorists preferred bayonets, such as the French, soldiers realized that actual bayonet fighting was rare in this era. French military authority Jacques Antoine de Guibert understood the issue in this way:

> Finally, I would add to this the exercise of the bayonet, consisting of putting it on the end of the barrel, replacing it in the sheath, and presenting it. I would not that, as is the fashion, the troops should appear in exercises, parades, [or] reviews with the bayonet; I would that it only be placed at the moment of combat or the simulated movements that represent it. The soldier is too familiarized with the bayonet and unnecessarily armed [with

---

49 William Dalrymple, *Tacticks* (Dublin: George Bonham, 1782), p.113.

it]. As such, he is accustomed to regarding it as a weapon without use. He used to estimate it as his last resource; a soldier, and a French soldier especially, said, 'I no longer have ammunition, but my bayonet remains with me.'....It is from the German infantry that we have taken the custom of always carrying the bayonet, and a singular thing it is that, since they have begun to always carry it, they have never used it.[50]

In short, during cavalry action, attacks on defensive works, and surprise attacks, troops often engaged in melee combat. These were the places for bayonets, not in the open field against other infantry. When cavalry troops were involved, melee combat was quite frequent. At Guilford Courthouse, William Washington's light dragoons savaged the British 2nd Guards Battalion.[51] One of these light dragoons, Peter Francisco, gave us a window in the visceral intensity of this type of combat: 'Colonel Washington, observing their maneuvering, made a charge upon them, in which charge he [Francisco] was wounded in the thigh by a bayonet, from the knee to the socket of the hip, and in the presence of many, he was seen to kill two men, besides making many other panes which were doubtless fatal to others.'[52]

At the Battle of Hohenfriedberg the Prussian Bayreuth Dragoon Regiment swept away a large portion of the Austrian army.[53] *Premierleutnant* Francois de Chasot, a friend of Frederick, gave his recollection of the attack:

> At the start we moved at a walk. We crossed several ditches one rank at a time, and on each occasion I made the leading rank halt of the far side so as to give the reward two ranks time to catch up. Then we broke into a trot, and finally into a full gallop, putting our heads down and running into the Austrian grenadiers. At first they stood bravely, and fired a salvo at twenty paces. After that, they were overthrown and cut down.[54]

When fighting other cavalry, it appears that the horses would seek intervals through the enemy formation, leading to brief moments of intense combat followed by manoeuvring. In moments of true melee combat, all order was lost and horses and men swirled around in individual combats. Even here, however, the cavalry melee was perhaps less effective than frequently believed. In the 1780s, military theorists studying the Prussian army recorded, 'We heard from some cavalry officers that when troops undertake a charge, almost always, one troop flees before melee is

---

50 Guibert and Abel (trans.), *Guibert's General Essay on Tactics*, p.58.
51 Lawrence E. Babits and Joshua B. Howard, *Long Obstinate and Bloody: The Battle of Guilford Courthouse* (Chapel Hill: University of North Carolina Press, 2009), p.160.
52 'Letter of Peter Francisco to the General Assembly, November 11, 1820', *The William and Mary Quarterly*, vol.13, no.4 (Apr. 1905), pp.218–219.
53 Christopher Duffy, *Frederick the Great: A Military Life* (London: Routledge & Kegan Paul, 1986), p.64.
54 Matthias Eberhard Kröger, Karl Theodor Gaedertz (ed.), *Friedrich der Grose und General Chasot* (Bremen: Müller, 1893), p.38.

joined, and the other gives pursuit.'[55] Georg Tempelhof, a veteran of the Seven Years War, reported:

> The strength of cavalry consists in its movement: it must have the ability to manoeuvre with speed. The shock or charge has no effect unless it happens in this way. Forgive me if I do not consider the cavalry's shock to be so decisive as it seems. In the 1762 campaign I observed Prussian cavalry charging superior Austrian horsemen. The result was that on both sides there were a few hundred wounded and prisoners. Not a single man lay dead on the battlefield.[56]

It is possible that Tempelhof, as an artillerist, underestimated the potency of cavalry. Horsemen, particularly in Europe, preferred to engage in hand-to-hand combat. Cavalry officers endlessly debated whether or not it was more effective to cut or thrust against the enemy, or whether straight or curved swords were more effective. Frederick II was once pressed on this issue: 'Speaking one day with his Majesty the King of Prussia, of this diversity of opinions, with regard to the edge or the point, he answered, "Kill your enemy with the one or the other, I will never bring you to an account with which you did it."'[57]

Having briefly addressed cavalry, we will now return to infantry, the primary focus of this work. In addition to an attack by cavalry, another example where troops might find themselves in melee combat was fighting over defended positions. When defending soldiers held fortified or prepared positions, melee combat could be fierce. Soldiers attached great psychological importance to their defensive works, and often tangled with enemy troops in melee combat in order to defend them. In discussing this issue, Christopher Duffy presents evidence that soldiers actually had difficulty understanding that they needed to use their bayonets in these types of assaults.[58] Even insignificant defensive positions could motivate defending troops to stand, such as the rail fence which the North Carolina troops sheltered behind at Guilford Courthouse in 1781. An advancing Hessian soldier, recalling the fighting over this obstacle, only commented, 'Colonel Du Buy at once ordered, 'Fix bayonets! March!' Before the enemy could reload, we changed against them with our bayonets. Everyone was bayoneted.'[59]

Obviously, this type of melee action includes siege warfare. During the Siege of Schweidnitz in 1762, Austrian *Premierleutnant* Waldhütter and 30 men of the Erzherzog Carl regiment spearheaded a successful sortie against the besieging Prussian forces. Franz Guasco, the fortress commandant, left this description of the sortie:

---

55 Honoré-Gabriel de Riqueti Mirabeau, *Système militaire de la Prusse* (London: Maradan, 1788). p.104
56 Georg Friedrich Tempelhof, *Geschichte des Siebenjährigen Krieges in Deutschland* (Berlin: Johann F. Unger, 1783), vol.1, p.68.
57 De Warnery, *Remarks on Cavalry by the Prussian Major General Warnery*, p.17
58 Duffy, *Military Experience in the Age of Reason*, pp.204–205.
59 Koch and Burgoyne, *The Battle of Guilford Courthouse and the Siege and Surrender at Yorktown*, p.7.

> Waldhütter and his troops jumped inside without hesitation and found the Prussians on their guard. Some of them opened fire, while some knelt on the floor and raised their muskets, the bayonets fixed to the muzzles. Our men flung themselves blindly among them, sabre in hand; some of them were skewered on the bayonets, but the rest set about the enemy and hacked them to pieces.[60]

Alexander Hamilton was briefly involved in bayonet fighting during the storming of redoubts nine and 10 at the Siege of Yorktown. Joseph Plumb Martin is silent on the exact nature of the fighting inside the redoubts.[61] Hamilton reported that men under his command suffered a number of 'bayonet wounds' in the course of the fighting.[62]

Troops often fought with bayonets during surprise attacks or 'massacres.' The American War of Independence produced a number of famous night attacks which led to bayonet fighting. Both of these, at Paoli in 1777 and Tappan in 1778, were extremely violent affairs, with many soldiers being killed both in and out of combat. At both Paoli and Tappan, British troops bayoneted Americans, and there is indeed evidence that Americans fought back with bayonets. This type of fighting was quite intense from a psychological perspective, and you often find descriptions of men, who, like Captain Sir James Baird, killed large numbers of enemy troops singlehandedly. In addition to fighting with bayonets, troops at Paoli and Tappan burned to death and were killed by firepower. Lieutenant Martin Hunter described the scene: 'the camp was immediately set on fire; the Light Infantry bayonetted every man they came up with... this, with the cries of the wounded, formed altogether the most dreadful scene I ever beheld. Every man that fired was immediately put to death.'[63]

During the Seven Years War, the Austrians managed to inflict similar damage on the Prussian army. At the Battle of Hochkirch in 1758, Austrian columns overran the Prussian camp before some Prussians were even awake. Johann Wilhelm von Archenholz, a Prussian veteran, described what it was like the be on the receiving end of such a swift-moving attack:

> It was dark, and confusion reigned supreme. What a sight for these warriors, almost like a night terror. The Austrians seemed to emerge from the earth, in the midst of the Prussian flags at the centre of camp! Several hundred men were killed before they could open their eyes, and others ran half-naked to their weapons. Only a few could reach them. Others laid ahold of whatever was closest to hand and began to fight.[64]

---

60   Quoted in Christopher Duffy, *By Force of Arms: The Austrian Army in the Seven Years War Volume 2*, p.374.
61   Joseph Plumb Martin, *A Narrative of some Adventures, Dangers, and Sufferings of a Revolutionary Soldier* (Hallowell: Glazier, Masters & Co, 1830), pp.170-171.
62   Alexander Hamilton and Henry Cabot Lodge (ed.), *The Works of Alexander Hamilton* (New York: Knickerbocker Press, 1904), vol.8, p.47.
63   Martin Hunter, *The Journal of General Sir Martin Hunter and some letters of his Wife Lady Hunter* (Edinburgh: The Edinburgh Press, 1894), p.31.
64   Johann Wilhelm von Archenholz, *Geschichte des Seibenjährigen Krieges in Deutschland* (Berlin: Haude und Spener, 1793), vol.1, p.279.

Here we have an example of true melee, with thousands of soldiers fighting and dying in close combat. In the confusion, troops searched for uniform details to determine friend from foe, like the metal or bearskin caps of grenadiers.[65] However, even here, it is important to note the ways in which this melee is unique. The Prussian army was surprised in camp and fought a battle of desperation because of the impossibility of escape. In the open, troops would have fled long before this point.

So, did troops in the open really not cross bayonets? Outside of the abovementioned categories, infantry forces actually crossing bayonets when one force had the option to flee seems to have been quite rare. Again, there are instances such as Culloden and Guilford Courthouse, but those remain more the exception than the rule. According to one French report, 68.8 percent of troops were wounded by small arms fire, 14.7 percent were wounded by artillery fire, and approximately 15 percent were wounded by swords and bayonets.[66] When we consider that swords were the cavalry's main form of engaging the enemy, these figures are impressive. J.F. Puysegur argued: 'Essentially, the firearm is the weapon that does the most damage and today more than ever. To be convinced of this, one must only go to the hospitals, and you will see how few men have been wounded by cold steel as opposed to the number firearms. My idea is not advanced lightly. It is founded on knowledge.'[67]

Because Puysegur was writing in the 1740s before the Seven Years War, his experience is even more telling. Other military theorists, such as David Dundas, recalled, '... infantry seldom mix with bayonets.'[68] Historians should carefully weigh the value of modern combat psychologists in understanding the fighting of the past. This was one of Christopher Duffy's principal arguments in *Military Experience of the Age of Reason*. With that caveat in mind, the reluctance of eighteenth-century fighting men to close with bayonets does seem to present a continuity with more modern soldiers. These eighteenth-century military observations match well with those from the twentieth century:

> The vast majority of soldiers who do approach bayonet range with the enemy use the butt of the weapon or any other available means to incapacitate the enemy rather than skewer him... when the bayonet is used, the close range results in a situation with enormous potential for psychological trauma... The resistance to killing with the bayonet is equal only to the enemy's horror at having this done to him. Thus, in bayonet charges one side or the other invariably flees before the actual crossing of bayonets occurs.[69]

---

65   See Christopher Duffy, *By Force of Arms*, p.374 and also Johann Wilhelm von Archenholz and Max von Duvernoy, *Geschichte des Seibenjährigen Krieges in Deutschland* (Leipzig: Amelangs, 1911), p.188.
66   André Corvisier, *L'armée française de la fin du XVIIe siècle au ministère de Choiseul : le soldat* (Paris: Presses universitaires de France, 1964), p.64.
67   Jacques François de Chastenet Puységur, *Art de La Guerre par principes et por regles* (Paris: Jombert, 1749), vol.1, p.227.
68   Dundas, *Principles of Military Movements Chiefly Applied to Infantry*, p.51.
69   David Grossman, *On Killing: The Psychological Cost of Learning to Kill in War and Society* (Boston: Bay Back Books, 2009), p.51.

This raises two points. First, troops preferred to engage their enemies with weapons other than the bayonet. Second, troops often fled when presented with an enemy's bayonet. On the first point, even in desperate circumstances, such as the battle of the third line at Guilford Courthouse, we see infantry attempting to load and fire *while in* melee combat. Captain John Smith of the 1st Maryland Regiment found himself in heavy melee combat against the British Guards but was shot in the head at an extremely close range (non-fatally by buckshot) by a soldier *who had just loaded*.[70] Therefore, there may be some truth to the idea of psychological prejudice against using bayonets, even in the eighteenth century. Lieutenant Colonel John Graves Simcoe described realistic small unit training which kept this principle in mind:

> …they were, particularly, trained to attack a supposed enemy, posted behind railing, the common position of the rebels; they were instructed not to fire, but to charge their bayonets with their muskets loaded, and, upon their arrival at the fence, each soldier to take his aim at their opponents, who were then supposed to have been driven from it; they were taught that, in the position of running, their bodies afforded a less and more uncertain mark to their antagonists, whose minds also must be perturbed by the rapidity of their approach…[71]

Here, we can see that when driving enemy from a position, infantry were trained to fire at fleeing men, rather than attempt to chase them down with bayonets. Now, turning to the second point, we do see frequent examples of troops actually fleeing as a result of being threatened with bayonets. At Kesselsdorf in 1745, Prince Henry of Prussia's Regiment, led in person by Prinz Moritz of Anhalt-Dessau, chased the enemy off the battlefield with this type of assault. An officer present at the scene recalled, 'Prinz Moritz flew in front of the Prince of Prussia Regiment and led it, sword in hand, and broke the Weissenfels Regiment with felled bayonets, and which then threw itself on the enemy second line, which also withdrew before it.'[72] In this attack, Prussian troops broke the enemy, not by an actual extended melee, but by the threat of it.

American officer and writer Otho Holland Williams described this type of attack perfectly at the Battle of Eutaw Springs on 8 September 1781. Williams, an officer advancing to the attack the British line with bayonets, recalled: 'If two lines on this occasion did not actually come to the mutual thrust of the bayonet, it must be acknowledged that no troops ever came nearer. They are said to have been so near that their bayonets clashed and officers sprang at each other with their swords before the enemy actually broke away.'[73]

This bayonet attack represents the pinnacle of American military training during the War of Independence. Of course, British bayonet attacks which forced

---

70 Babits and Howard, *Long Obstinate and Bloody: The Battle of Guilford Courthouse*, p.160
71 John Graves Simcoe, *Simcoe's Military Journal: A History of the Operations of a Partisan Corps* (New York: Bartlett and Welford, 1844), p.98.
72 Anon., *Sammlung ungedruckter Nachrichten*, vol.1, p.431.
73 'Battle of Eutaw: Account furnished by Otho Williams', in Robert Wilson Gibbes, *Documentary History of the American Revolution* (Columbia: Banner Press, 1853), vol.3, p.150.

the American troops to withdraw were much more common. American soldier Cornielus Sullivan at the Battle of White Plains in 1776 recalled, 'The British advanced at a charge bayonets and approached so near that [he] could see the buckles on their shoes. The Americans then retreated.'[74] Indeed, as the Marquis de Lafayette recalled, American troops were even hesitant to fix bayonets as late as the Battle of Brandywine, when: 'He dismounted and did his utmost to make the men charge with fixed bayonets. The Frenchmen personally attached their bayonets for them, and Lafayette pushed them in the back to make them charge. But the Americans are not suited for this type of combat, and never wanted to take it up.'[75]

Obviously, melee occurred in the eighteenth century. However, it was not a common occurrence in the open when infantry fought one another and seems to have become less prevalent over the course of the era.

Delivering Intimidation: The Use of Column Assault Formations

The French chevalier de Folard ignited a firestorm of controversy over the use of the column as a tactical formation in the 1730s, and this continued right down to the era of the French Revolution. Folard believed that with a superior amount of mass, a column could break through the thin linear order which reigned supreme on eighteenth-century battlefields. The French attack at Rossbach is a well-known example of an impromptu columnar assault.

It is usually asserted that attacks by columns were not a feature of mid-eighteenth-century warfare, but only appeared with the advent of Revolutionary or Napoleonic warfare. There is no question: during this later period, troops used a greater variety of column formations at the battalion, regimental, and division level. However, this chapter demonstrates that the Austrians did attack in columnar formations during the Seven Years War. These formations were not only 'march formations' which took the unit to the battlefield, but were also used within musket range of the enemy troops.

This chapter does not look at the most-often cited example of an attack in column during the Seven Years War: the abortive French infantry attack at Rossbach. This is clearly an unintentional use of the column, born out of dire necessity. Likewise, the current author will not tap into the extensive theoretical debate regarding the use of columns from Folard on. Rather, this chapter looks at battles where the commanders made a conscious decision to engage enemy forces, whether in column or not. At the battles of Moys, Hochkirch, Maxen, and Landeshut, the Austrians used a variety of successive linear and column formations in order to approach and attack the enemy positions.[76] At Adelsbach, the Prussians did the same. In each of these situations, circumstances and the terrain conspired to make attacking in a deep formation the most effective way of combating the enemy. At Moys, Hochkirch, and Maxen, the

---

74   NARA: M804, Pension Application of Cornelius Sullivan S. 1258
75   Marquis de Lafayette and Stanley J. Idzerda, *Lafayette in the Age of the American Revolution: Selection Letters and Papers* (Ithaca: Cornell University Press, 1977), vol.1, p.84.
76   In using the language of a 'successive linear' attack, I have followed the convention of Paddy Griffith, *Battle Tactics of the Civil War* (Ramsburg: Crowood Press, 2014) pp.151–152.

Austrians attacked on a battalion frontage – what we might call battalion columns, or successive linear waves. At Landeshut, a single Austrian grenadier battalion attacked in a column of companies. At Adelsbach, the terrain forced the Prussian troops to approach the enemy position in a column.

At Moys in 1757, Christopher Duffy has clearly demonstrated that the Austrians employed a successive linear attack.[77] The Austrians arrayed their battalions in seven 'columns' of three battalions each, separated by 100 yards. Within each 'column' the battalions had 200-yard intervals between them. This allowed for flexibility, as the orders explained: 'if a first-line battalion suffers heavy losses, or falls into disorder, we will file it off to the left or right, and replace it with the battalion behind.'[78] This columnar, or successive linear assault, would form a model for the Austrians during the war, as they attacked using columns with a battalion frontage again at Hochkirch and Maxen.

At Hochkirch in 1758, the novel method of approach and attack meant that different officers had immediate tactical control of sectors of the battlefield; thus, some of the attacking 'columns' formed into line of battle earlier, while others persisted in a columnar formation.[79] The Austrian veteran Cognazzio asserts that this was altogether too much to ask of the troops, and that his component division within his battalion was a mongrel force of collected men: 'Grenadiers, Fusiliers, Hungarians, and Germans... [I] placed them together in rank and file, and brought the line into being.'[80] Cognazzio asserts that in the heat of the fight, flexibility was the only thing that allowed for the creation of a 'well-closed line.'[81] So, at least in some instances, it seems that the columns of Hochkirch were intended to deliver men to the area of action, rather than a formation by which to actually conduct an attack.

At Maxen in 1759, the situation is rather different. Here, the depth of the 'battalion columns' of the Austrians was significantly increased to 12 battalions. Led to the attack by the grenadiers of the army, two Austrian battalion columns approached the enemy positions. The Austrian official report cites that they were greatly supported by an artillery bombardment, and seeing that: 'Such a swift, sustained, and well-placed fire had caused great damage to the enemy lines, and that they were beginning to waver, the assault was allowed to go forward. It happened that the infantry were in battalion columns.'[82] The same source continues, asserting that the battalion columns were not formed into a wider battle line until the Prussian position on the heights was broken.[83]

---

77 Duffy, *Prussia's Glory*, p.97.
78 Quoted in Duffy, *Prussia's Glory*, p.97.
79 Duffy, *By Force of Arms*, pp.129–143.
80 Jakob Cognazzio, *Geständnisse eines Oesterreichischen Veterans* (Breslau: Gottlieb Löwe, 1790), vol.3, p.47.
81 Jakob Cognazzio, *Geständnisse eines Oesterreichischen Veterans* (Breslau: Gottlieb Löwe, 1790), vol.3, p.48.
82 J. G. Tielcke, *Beytraege zur kriegs-kunst und Geschichte des Krieges*, (Freyberg: Barthel,1775), vol.1, p.30.
83 J. G. Tielcke, *Beytraege zur kriegs-kunst und Geschichte des Krieges*, (Freyberg: Barthel,1775), vol.1, p.31.

Historians should use visual sources, even those painted closely after events, with extreme care when reconstructing battles. *The Attack at Maxen,* painted immediately after the battle by Franz Paul Findenigg, displays some features worthy of note. First, Findenigg correctly identifies the first two battalions approaching the Prussians as grenadiers (they have peaked caps, and carry no flags, while the other battalions all carry flags and wear cocked hats). Findenigg also depicts the action of the battery disrupting the Prussians, as well as Austrian battalions in a 'successive line' or 'battalion column' formation, stacked several units deep. The individual battalions of the column seem much closer than the guidelines of the attack at Moys, perhaps supporting Christopher Duffy's assertion that 'Austrian column[s] of assault' were formed in a dense closed-up formation.[84] We should not put too much weight on this visual evidence. However, while presenting the same issues as other visual sources, Findenigg's painting seems to support the idea that the Austrians made their initial breakthrough of the Prussian line using this formation.

At Landeshut in 1760, we find something rather different. Here, two grenadier battalions led the attack on Prussian fixed positions on the Mummelberg and Buchberg. The grenadier battalion of *Major* de Vins employed a column of companies for this assault.[85] Having taken these two positions and been returned to order, a larger force of infantry now combined into two 'columns' and launched an 'assault of columns' against the Prussian position on the Kirchberg. As opposed to a column of companies, this attack, especially considering the way it is depicted on a map printed after the war in 1790, (three lines), was likely a successive linear wave attack as at Moys.[86]

The Prussian use of columns in the attack at Adelsbach on 6 July 1762 appears to have been largely unintentional. Attempting to get at the Austrian position, the Prussians had to march down a valley, through Ober Adelsbach, over the stream, and back up a valley to the heights where the Austrians were waiting for them. As a result, they were unable to properly form for the attack, and came on in some sort of marching column, likely of open platoons. The Prince de Ligne noted that the Prussians were marching to the attack '*dû défilér*', indicating a formation narrower than a line.[87] Upon reaching the height, however, they attempted form a more traditional battle line.[88] This was not, therefore, an intentional attack in a column, but one mandated by the terrain.

Thus, during the Seven Years War, the Austrian army did attack in a columnar formation. These attacks were less varied, less coordinated, and more ad-hoc than later Revolutionary and Napoleonic attacks in column. Despite this, the evidence is clear: the Austrians did indeed innovate with alternative linear and columnar attacks during the Seven Years War, not just in theory, but actually on the battlefield. This may have been unintentional, a feature of the novel Austrian grand tactics of

---

84 Duffy, *Instrument of War*, p.405.
85 Duffy, *By Force of Arms*, p.233.
86 Johann Christian Jaeger, *Plans von Zwey un Vierzig Haupt Schlachten, Treffen, und Belagerungen* (Frankfurt: Jaegerishen Buchhandlung, 1790), pp.124–125.
87 Prince de Ligne, *Melanges militaires, litteraires, et sentimentaires* (Vienna: Publisher Unknown, 1796), vol.16, p.124.
88 Gabriel Nicolaus Raspe, *Plan von der Affaire ... am 6. July 1762 bey Adelsbach* (Nürnberg: Raspischen Handlung, 1763), legend entry E.

the time. Innovation is sometimes unintentional. By the time the Seven Years War had ended, the Austrians had attacked in successive linear waves, columns of battalions, and a battalion column of companies.

Conclusion

Between 1733 and 1783, troops from a variety of states across military Europe took cover from enemy fire, moved at speed on the battlefield, tried to avoid hand to hand combat, and experimented with novel forms of infantry attack. Taken together with their predilection for firing at longer range and firing without orders, and the possibility of fighting as skirmishers, these infantrymen experienced many of the features of modern infantry combat while remaining in an early-modern world. Napoleonic historians have long attempted to emphasize a radical tactical 'break' between the eighteenth-century wars and the Napoleonic period particularly outside of Britain in states like France and Prussia.[89] At least in the realm infantrymen in battle at the tactical level, eighteenth-century infantry combat was not rigidly confined to 'unthinking' or 'robotic' linear infantry warfare. These troops fought in villages, laid down under fire, ran on the battlefield, and dealt with the psychological hardships of close combat.

Having examined the thematic arc of infantrymen in battle between 1733 and 1783, the book will now end as it began, with an examination of a pair of battles from the end of the period. The two battles of Germantown (1777) and Eutaw Springs (1781) are remarkable examples, illustrating many of themes which the book has developed up to this point. Like at Parma and Guastalla in 1734, these battles display the full range of possibilities for troops in combat, as they assaulted enemy positions, took cover from enemy fire by utilizing micro-tactical terrain, and fired on their opponents from a psychological need rather than orders from their officers. If Parma and Guastalla began a unique period of infantry battle, Germantown and Eutaw Springs show the ways that over a 50-year period, many of these themes endured despite the intervening Prussomania.

---

[89] See, for example: Michael V. Leggiere, 'Napoleon and the Strategy of the Single Point,' in Hal Brands, *The New Makers of Modern Strategy* (Princeton: Princeton University Press, 2023) pp.320–321; James A. Arnold, *October Triumph: Napoleon's Invasion of Germany, 1806 Jena and Auerstädt* (Lexington: Napoleon Books, 2020) pp.10–31. John A. Lynn, *The Bayonets of the Republic: Motivation and Tactics in the Army of Revolutionary France, 1791-94* (Champaign: University of Illinois Press, 1984), pp.300–332.

9

North America, 1777 and 1781

> *Soon after this as we were descending a hill through an orchard, a party of the enemy who were entrenched behind a bank and fence, rose and fired upon us.*[1]
>
> American Sergeant Nathaniel Root, 1777

In the early hours of a warm day, troops of a doctrinally defensive army launch a counterattack, hoping to avenge themselves against a traditionally aggressive opponent. As the battle lines close, a firefight erupts between the leading elements of the attacking army and their enemies, dashing their commander's hopes for a quick attack with the bayonet. After applying pressure, the attacking army manages to break through the first line of enemy opposition – victory seems at hand.

Their opponents, so used to seizing the offensive themselves, have an answer to this. Seizing on a key piece of terrain, the retreating forces garrison a multi-story stone or brick building, occupying it as a strongpoint. This takes the attackers by surprise. They are elated to have sent their often-successful opponents into retreat but are in a quandary as to how to deal with the new challenge of the strongpoint. The battle continues for some time around and beyond the strongpoint before a shortage of ammunition forces the attackers to quit the battlefield. Victory, seemingly so close, has eluded the novice attackers.

This pattern of battle characterizes many eighteenth-century combats, not least the Battles of Parma and Guastalla which began this book. In many ways, those battles started a 50-year experience of combat stretching from the War of Polish Succession to the American War of Independence. In this final chapter, we will examine a pair of battles from the end of that date range in the American War – Germantown and Eutaw Springs. Both provide an important window into the experience of battle for infantrymen, the primary combatants during this conflict.

It is important to stress that these battles were not typical of the fighting during the American War of Independence. In the majority of battles during the American War, British infantry, deployed in a two-rank, open-order line of battle, and launched swift-moving attacks which drove the Americans from the field. Occasionally, the Americans were able to successfully oppose this flexible and aggressive doctrine. More frequently, they were able to inflict damage on the British during their advance

---

1   Nathaniel Root, 'The Battle of Princeton,' *Pennsylvania Magazine of History and Biography*, vol.20 (1896) p.517.

even if defeated in the final outcome. Relatively rarely, American commanders seized the offensive themselves, launching counterattacks that turned the tables on the aggressive British infantry. This chapter explores two such battles from the American War.

Unlike Parma and Guastalla, there is a large literature on Germantown and Eutaw Springs, part of an even larger literature on the American War of Independence in general. In covering these two battles in this single chapter, my goal is not to provide a definitive battle study of either, which has already been done.[2] Instead, the goal of this chapter is to draw out parallels and themes from these battles, in order to contrast their experience with the two clashes previously discussed at Parma and Guastalla. For the American, British, and German-speaking soldiers who took part, these battles were an important reminder of the importance of key terrain, the preference of soldiers for firepower over hand-to-hand combat, and the vital task of keeping soldiers well supplied with ammunition. Like Parma and Guastalla, Germantown and Eutaw Springs show the ways that eighteenth-century warfare could be surprisingly modern.

The American War of Independence was the military struggle associated with the American Revolution, lasting from 1775 to 1783. In this struggle, a tactically aggressive and proficient British Army attempted to hunt down, bring to battle, and destroy the American Continental Army. Matthew H. Spring's 2008 study *With Zeal and With Bayonets Only* has redefined historians' understandings of the British Army which fought this war. Deployed in a thin two-rank and open-order line of battle, the British jogged into combat, trying to drive the American forces off the battlefield in a series of quick charges.[3] The literature still needs a comparable study on the reality of American tactical practices.

As historians have increasingly noted since 2000 this conflict took on global dimensions comparable to the War of Austrian Succession and Seven Years War, especially after 1777. The French were immediately involved, supporting the United States with financial and logistical support. Eventually, the French, Spanish, and Dutch also fought related and independent wars with Britain. Fighting raged in the Spanish and British colonies on the Gulf of Mexico, the Caribbean, in European coastal waters, in Gibraltar and the Mediterranean, on the subcontinent of India, and in the Indian Ocean. Austria, Prussia, Russia, and Saxony engaged in a politically unrelated but concurrent struggle, the War of Bavarian Succession, from 1778–1779. Eventually, French forces would be deployed directly to America, fighting alongside the Continental Army, and helped to assure the victory of American forces over the British.

This war contained a plethora of army-level commanders, so the chapter will only introduce those directly involved in the battles described below. On the American

---

2   For Germantown, see: Thomas McGuire, *The Philadelphia Campaign, Volume II: Germantown and the Roads to Valley Forge* (Mechanicsburg: Stackpole, 2007). For Eutaw Springs, see: Robert Dunkerly and Irene Boland, *Eutaw Springs: The Final Battle of the American Revolution's Southern Campaign* (Columbia: University of South Carolina Press, 2017).

3   Matthew H. Spring, *With Zeal and With Bayonets Only: The British American on Campaign in North America,* (Norman: University of Oklahoma Press, 2008).

side, the most successful was George Washington: a commander who won the war by surviving rather than achieving decisive battlefield victories. Washington managed to keep his army together in the face of adversity, and was handed a number of stinging defeats by more tactically proficient British forces, particularly at Long Island (1776) and Brandywine (1777). Although not the most adroit battle manager, he did score impressive victories during surprise counteroffensives at Trenton (1776) and Princeton (1777) and fought an already retreating British force to a standstill at Monmouth Courthouse (1778).[4]

Washington's most successful subordinate commander, Nathanael Greene, fought in the battles of the Northern Campaigns under Washington. Greene showed tactical talent at Springfield (1780) and, after Horatio Gates's failure at Camden, was sent south to take command of the Southern Department of the Continental Army. His record in the south was mixed. His subordinate, Daniel Morgan, won a small but famous victory at Cowpens (1781) while Greene was defeated at Guilford Court House (1781) two months later. The defeat at Guilford Court House badly damaged British commander Charles Cornwallis's army, giving Greene freedom to operate against Camden, resulting in another defeat at Hobkirk's Hill. In September of 1781, Greene fought his last major battle at Eutaw Springs, which resulted in an immediate defeat for his army, but forced his British opponents to withdraw to Charleston.

For much of the early war, British forces in North America largely fell under the leadership of Sir William Howe. Howe was a capable battlefield manager who outmanoeuvred Washington tactically and operationally but failed to deliver a war-winning stroke. He defeated Washington at Long Island, captured thousands of Americans at Fort Washington (1776) and planned an invasion of Philadelphia that culminated in the capture of the American capital in 1777. He was unable to decisively crush the Continental Army but outflanked and badly defeated Washington at Brandywine on 11 September 1777. Frustrated with a lack of support from the British government, he resigned in October of 1777.

The British forces directly opposing Nathanael Greene in the south after 1780 were led by three principal commanders: Charles Cornwallis, Francis Rawdon-Hastings (Lord Rawdon), and Alexander Stewart. Cornwallis left the Carolinas operational area after his pyrrhic victory at Guilford, Lord Rawdon was taken ill after his victory at Hobkirk's Hill and the relief of Ninety-Six, and Lieutenant Colonel Alexander Stewart was left in control of the only field army capable of opposing Greene in the Carolinas. Stewart was a career line officer, having served in the 37th Regiment during the Seven Years War. He eventually transferred to the 3rd Regiment ('The Buffs') and was promoted to Lieutenant Colonel in 1780. As a Lieutenant Colonel, he was left to try and oppose Major General Greene's larger American force from encircling Charleston, South Carolina. With the commanders described, the chapter will now turn to describing the first of the two battles in the chapter: Germantown.

---

4   For a recent military biography of Washington, see: Edward Lengel, *General George Washington: A Military Life* (New York: Random House, 2007).

Germantown, 4 October 1777

After the army-wide defeat at Brandywine on 11 September, and the losses suffered by Anthony Wayne's Pennsylvania troops at Paoli on the 20th, Washington decided to withdraw from Philadelphia, opening the way for William Howe's troops to occupy the city.

Although remaining on the defensive for much of the war, Continental troops occasionally marched to the attack, utilizing new European ideas on how best to dispense troops to attack enemy positions. During much of the eighteenth century until the War of Austrian Succession/War of Jenkins Ear (1739–1748), it was common for attacking forces to approach enemy positions in marching columns, and then deploy to form a unitary line of battle, which might consist of up to three lines of infantry deployed centrally, with cavalry on the wings of this formation. During the Seven Years War, the Austrians pioneered a different style of attack, by which multiple columns would advance on different roads and attack different points of the enemy position simultaneous, acting not as one unified body but independent bodies of troops in concert. This type of operational plan could surround a numerically smaller enemy or attack at distant points simultaneously, tying down enemy reserves. In the Battle of Hochkirch on 14 October 1758, the Austrians introduced this method of attack, and it was quickly copied by the Prussians and other European forces. Christopher Duffy, an authority on the Austrian army during the Seven Years War, describing the development of this method of offensive warfare, writes:

> It can hardly be emphasized strongly enough that the form of attack was entirely novel, namely by means of independent, converging columns, a form of grand tactics which influenced the Austrian way of making war until the 1790s. [Franz Moritz von] Lacy as chief of staff devised the scheme as a whole, though an important contribution was made by Colonel Charles Amadei, who writes that he had been, 'entrusted with the chief attack, and in a conference or meeting beforehand I proposed to assault in columns, as giving a greater chance of success.'[5]

With all of Duffy's praise for the novel nature of this style of attack, it is worth mentioning that Ferdinand of Brunswick attempted a vaguely similar attack at Krefeld in June of 1758, dividing his force attacking the French into four columns. If Lacy and Ferdinand of Brunswick developed this type of attack in 1750s Europe, George Washington appears to have seized upon it in 1770s North America. Washington favoured this type of attack, using it at Trenton in 1776 and Germantown in 1777. At Trenton, Washington had hoped to cross an ice-filled river at three points, personally with Greene and Sullivan at McConkey's Ferry, while General James Ewing crossed near Trenton itself, and Colonel John Cadwalader

---

5 Christopher Duffy, *By Force of Arms: The Austrian Army in the Seven Years War Volume 2*, p.131.

crossed at Burlington.⁶ Ice flows and jams prevented Ewing and Cadawalder from crossing the river, but Washington was still able to surround the Hessian garrison at Trenton, marching his force in two columns from McConkey's Ferry to the town. The Trenton Campaign, then, is an early window into Washington's model for offensive grand tactics: one that tried to employ wide-ranging columns, marching in concert towards a single objective.

Just 10 days shy of the nineteenth anniversary of the above noted use of the attack in Lusatia, Washington launched a similar attack in Pennsylvania at the Battle of Germantown on 4 October 1777. Both at Hochkirch in 1758 and at Germantown in 1777, a numerically superior army which had struggled to defeat an aggressive tactically superior enemy took the offensive. This force began a complex early-morning attack employing converging columns in order to quickly deliver large numbers of troops into combat across the battlefield. Washington's 'General Orders for Attacking Germantown,' given on 3 October 1777, are a classic example of a European *Disposition,* where a commander gave complex and detailed instructions to each body of troops preparing to march for battle.⁷ In this document, Washington instructs each of his division commanders regarding their tasks and, hoping to completely envelope the enemy position, attacking both of their flanks. After giving each senior officer and their men a task, Washington closes, 'each column to make their disposition so as to attack the pickets in their respective routs, precisely at five o'clock [the next morning] with charged bayonets and without firing, and the columns to move on to the attack as soon as possible.' Washington, continuing his instructions, notes, 'The columns to endeavor to get within two miles of the enemy's pickets on their respective routs by two o'clock and there halt 'till four and make the disposition for attacking the pickets at the time above mentioned.'⁸

Continental officers noted the unusual method of march, and one captain from Delaware stated, 'The whole army marched in different divisions by different roads.'⁹ Like the Austrians at Hochkirch, Washington's plan called for a simultaneous assault at five in the morning, the precise time of the Austrian attack in 1758. Like the Austrians, Washington knew that his men would have to march all night in order to reach their stepping-off points. He hoped that they would be in position by two in the morning. Unlike the Austrians, Washington failed to account for the different lengths of time it would take the various columns to reach their appointed time. Thus, the results for Washington were somewhat less complete than those of Lacy and Daun at Hochkirch, but the Continental troops still had the satisfaction of driving their enemies before them in attack by a main Continental Army on a main British army.¹⁰

---

6   For an excellent summary of the operations during this phase of the Trenton Campaign, see: Hackett Fischer, *Washington's Crossing*, pp.208–220.
7   See Duffy, *The Military Experience in the Age of Reason*, pp.192–193.
8   LOC: Washington Papers, General Orders for Attacking Germantown, 3 October 1777.
9   Enoch Anderson, *Personal recollections of Captain Enoch Anderson* (Wilmington: Historical Society of Delaware, 1896), p.44.
10   For complete assessments of Hochkirch and Germantown, respectively, see: Duffy, *By Force of Arms: The Austrian Army in the Seven Years War Volume 2*, pp.129–148, and

182 INFANTRY IN BATTLE, 1733–1783

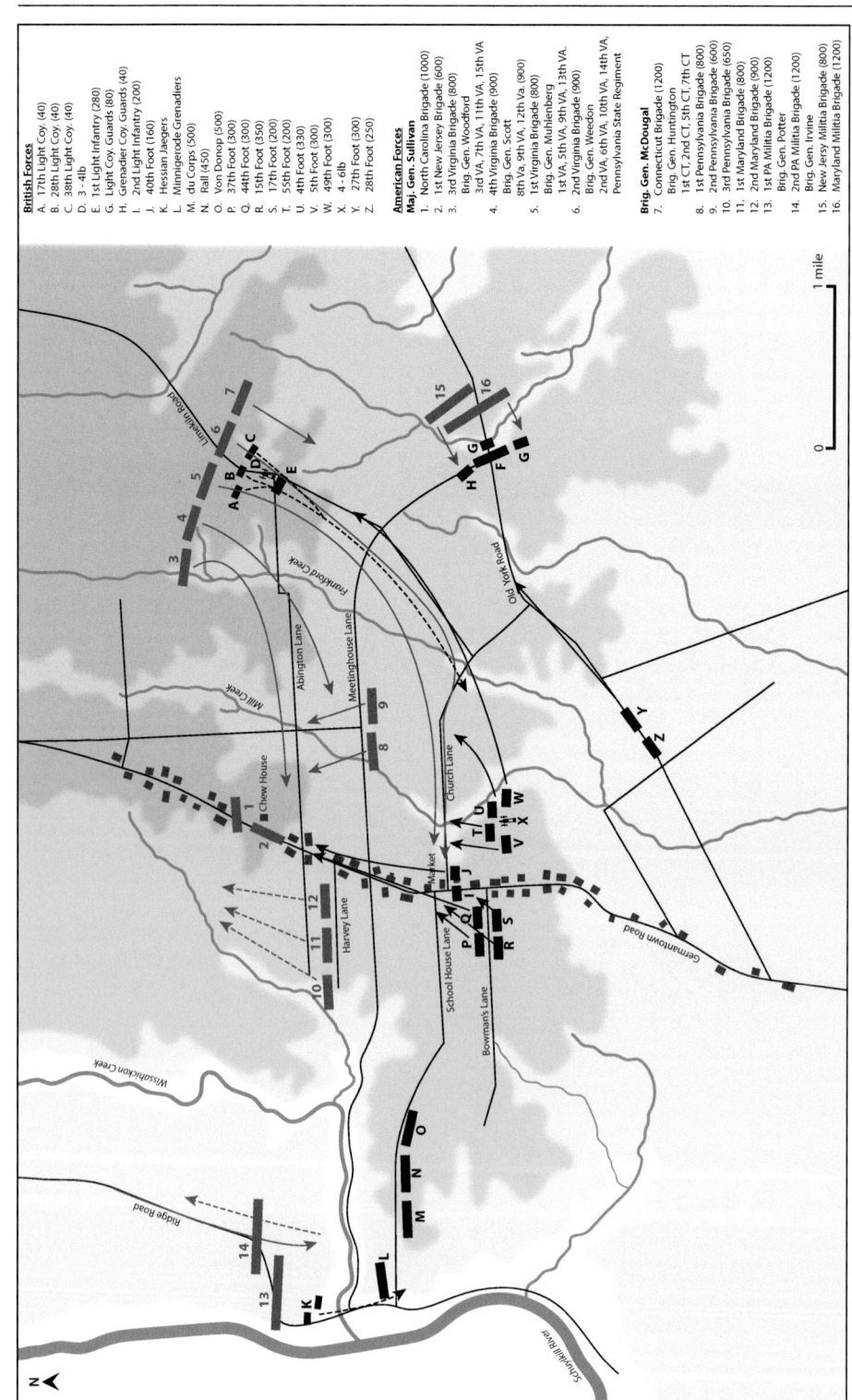

The Battle of Germantown 4 October 1777

Washington elected to attack Germantown on 4 October 1777, with an ambitious plan. A four-pronged attack, with militia composing the majority of the far-right and left-attacks, and continentals under Generals John Sullivan and Nathanael Greene in the centre-right and centre-left attacks, respectively. Eleven thousand American Continentals and militia squared off against the approximately 8,000 British and Hessian troops not engaged in the siege of Mud Island.[11]

The American attack on the far-right was mainly a holding or pinning action against the Hessian forces on the British left, while the main effort was made against the British right. General John Sullivan described the rationale for this:

> The reason of our sending so many troops to attack their right was because it was supposed that if this wing of the enemy could be forced their army must have been pushed into the Sculkill [Schuylkill River] or have been compelled to surrender. Therefore two thirds of the army at least were detached to oppose the enemy's right. The attack was to begin on all quarters at day break.[12]

Washington's general orders for the attack were detailed, covering topics from which units were to have attached artillery, to what sort of personal items the men should leave at camp or take with them on the march. Leaving the camp at six in the evening on 3 October, Washington wanted his men to 'endeavor to get within two miles of the enemy's picket on their respective routs by two o'clock and there halt till four and make the disposition for attacking the pickets.'[13] Washington then instructed 'each column to make their disposition so as to attack the pickets in their respective routs, precisely at five o'clock, with charged bayonets and without firing.'[14] In his general orders given the same day, Washington painfully reminded his men that they were being outdone by the northern American Army under General Horatio Gates at Freeman's Farm near Saratoga:

> ...in a capital action, the left wing only of General Gates' army maintained its ground, against the main body of the enemy our troops behaving with the highest spirit and bravery... This surely must animate every man, under the General's immediate command – This army – the main American army – will certainly not suffer itself to out done by their northern brethren – they will never endure such disgrace[.][15]

With this example of success ringing in their ears, Washington's columns moved out to attack the British near Germantown. Joseph Plumb Martin, marching with the

---

Thomas J. McGuire, *The Philadelphia Campaign: Germantown and the Roads to Valley Forge* (Mechanicsburg: Stackpole Books, 2007), vol.2.
11 McGuire *The Philadelphia Campaign,* vol.2, pp.48–49.
12 Otis G. Hammond, *Letters and Papers of Major-General John Sullivan, Continental Army* (Concord: New Hampshire Historical Society, 1930), vol.1, p.543.
13 LOC: Washington Papers, General Orders for Attacking Germantown, 3 October 1777.
14 LOC: Washington Papers, General Orders for Attacking Germantown, 3 October 1777.
15 LOC: Washington Papers, General Orders for Attacking Germantown, 3 October 1777.

8th Connecticut Regiment, remembered, 'Early in the evening we marched in the direction of Philadelphia, we naturally concluded there was something serious in the wind.'[16] At Germantown, the natural smoke-filled haziness of eighteenth-century battlefields was made much worse a blanket of fog. Lieutenant Martin Hunter with the 2nd Battalion of Light Infantry recalled, 'it was a very thick, foggy morning and so dark we could not see a hundred yards before us.'[17] Timothy Pickering, an aide at Washington's side, recalled, 'the haziness of the air, and its increased obscurity, from the burning of so much powder.'[18] After the battle, Washington referred to an 'unfortunate fog, joined with the smoke' which had prevented more success.[19] General Sullivan remembered, 'the misfortunes of this day were principally owing to a thick fog which being rendered still more so by the smoke of the cannon and musketry.'[20] Private Martin remembered, 'We marched slowly all night; in the morning there was a low vapour lying on the land which made it very difficult to distinguish objects at any considerable distance.'[21] Conclusively, then, even before the shooting started, it was difficult to see.

In the darkness and fog, not all American columns reached their intended starting positions, but at daybreak, Sullivan began the attack. Lieutenant Hunter, whose British light infantry unit was the first to be attacked, recalled, 'On the first shots being fired at our piquet the battalion was out and under arms in a minute.'[22] This statement, in and of itself, implies that Washington's orders to attack without firing had already been disobeyed. Throughout the day at Germantown, Continental officers would try to follow the spirit of Washington's desire to attack with charged bayonets, but like most eighteenth-century battles, Germantown was a firefight. The threat of bayonets, however, was still intimidating, and Hunter recalled of his unit, which had just bayonetted American General Anthony Wayne's men at Paoli: 'so much had they in recollection [of] Wayne's affair that many of them rushed out at the back part of the huts.'[23] The threat of American bayonets was very much on the light infantrymen's mind, and after the battle, they discussed the lucky decision to give up quarters in the nearby village of Beggarstown. Martin Hunter decided, 'It was a very fortunate circumstance for us that we had changed our quarters two days before from the houses at Beggarstown to wigwams outside the town, for I am certain, had we been quartered in the town the morning we were attacked, we should all have been bayonetted.'[24] The threat of bayonet attacks, despite the relative rarity of melee fighting, exercised a powerful influence over the minds of soldiers.

---

16 Joseph Plumb Martin, *A Narrative of some Adventures, Dangers, and Sufferings of a Revolutionary Soldier* (Hallowell: Glazier, Masters & Co, 1830), p.53.
17 Hunter, *The Journal of General Sir Martin Hunter*, p.33.
18 Timothy Pickering, 'Battle of Germantown', in *The North American Review*, vol.23 (1826), p.427.
19 LOC: Washington Papers, General Orders 5 October 1777.
20 Hammond, *Letters and Papers of Major-General John Sullivan*, vol.1, p.547.
21 Martin, *A Narrative of some Adventures*, p.53.
22 Hunter, *The Journal of General Sir Martin Hunter*, p.33.
23 Hunter, *The Journal of General Sir Martin Hunter*, p.33.
24 Hunter, *The Journal of General Sir Martin Hunter*, p.33.

Despite the firing, Continental officers did attempt to live up to Washington's orders. General Sullivan recalled 'we drove their left wing near three miles great part of the time, shouldered arms and charged bayonets.'[25] In another letter, he recalled, 'our men being ordered to march up with shouldered arms they obeyed without hesitation and the enemy retired.'[26] North Carolina troops reported, 'Our soldiers behaved with great resolution (and some pushed bayonettes).'[27] Major General Adam Stephen, who would be court-martialled for a friendly-fire incident which occurred during the fighting, recalled that in the opening stages of the battle, the troops 'pushed the enemy so closely, that I called to them, 'give them the bayonet'– upon hearing this, the enemy officers on horseback rode their rear out of sight; many of their men running after them'.[28] Once again, Stephen's story tells of the psychological impact that the threat of bayonet fighting could produce, rather than actually describing troops in hand-to-hand combat.

Describing the situation on the ground, Joseph Plumb Martin recalled officers pleading with men, 'not to fire till we could see the buttons on their clothes.'[29] Hunter vividly recalled American troops advancing to cheers: 'We heard a loud cry of 'Have at the Bloodhounds! [his unit's nickname] Revenge Wayne's affair!' and immediately [they] fired a volley. We gave them another in return.'[30] For all the desire to emulate the British attacks with the bayonet, firepower was the primary tool in the Continental infantryman's arsenal.

As the American forces advanced, the British began to retire. This was a new experience for many British soldiers, as the Battle of Germantown was the first action of the war where the main American army had taken the tactical offensive against the main British Army. Sullivan recalled, 'When our men first made the attack we had the most pleasing prospect before us… they were routed, pursued, and charged with fury.'[31] The situation was as pleasing for the Americans as it was distressing for the British. Martin Hunter recalled, 'This was the first time we had ever retreated from the Americans, and it was with great difficulty that we could prevail on the men to obey our orders.'[32] Now the fighting became protracted, as the British searched for every piece of tactical terrain to delay the American advance.

The American commander directly responsible for the heaviest fighting along the Germantown high street to Philadelphia, Sullivan recalled the defensive formula the British employed as they withdrew: 'Though the enemy were routed yet they took advantage of every yard, house, and hedge in their retreat [and] kept up an incessant fire through the whole pursuit.'[33] Sullivan insisted, 'however, they made a stand at

---

25 Hammond, *Letters and Papers of Major-General John Sullivan*, vol.1, p.576.
26 Hammond, *Letters and Papers of Major-General John Sullivan*, vol.1, p.545.
27 North Carolina Department of Archives and History: Executive Letter Book, John Penn and Cornelius Harnett to Richard Caswell, 20 October 1777, p.659.
28 LOC: Washington Papers, Adam Stephen to George Washington, 9 October 1777.
29 Martin, *A Narrative of some Adventures*, p.53.
30 Hunter, *The Journal of General Sir Martin Hunter*, p.33.
31 Hammond, *Letters and Papers of Major-General John Sullivan*, vol.1, p.567.
32 Hunter, *The Journal of General Sir Martin Hunter*, p.34.
33 Hammond, *Letters and Papers of Major-General John Sullivan*, vol.1, p.545.

every fence, wall, and ditch they passed, which were numerous.'³⁴ In other correspondence he remembered, 'the enemy made a stand at every wall and hedge and fence.'³⁵ This assertion is borne out by men in the ranks like Joseph Plumb Martin, who remembered facing down British infantry at Germantown: 'we saw a body of enemy drawn up behind a rail fence on our right flank; we immediately formed in line and advanced upon them.'³⁶

By now, the fog of battle had become fully apparent. An American soldier from Washington's camp reported to the *Newport Gazette*:

> ...it being a foggy morning the smoke and fire of cannon and musketry, the smoke of several fields of stubble, hay, and other combustibles... combined, made such a midnight darkness that great part of the time there was no discovering friend from foe but by the direction of the shot, and no other object but the flash of the gun.³⁷

Martin confirms this anecdote, noting that the enemy 'hid their clothes in fire and smoke.'³⁸ By this point in the battle Washington noted, 'Among other misfortunes that attended us, was a hazy atmosphere without a breath of air, so that the smoke of our artillery and small arms often prevented us from seeing thirty yards; and this not for an instant but of long continuance'.³⁹ Thus, the fight at Germantown had become a confused American advance, with the British contesting every viable piece of tactical terrain in the built-up area around Germantown. The battlefield had become so blanketed by fog and smoke that many troops were firing at enemy muzzle flashes as their main point of reference.

In this confusion, British troops identified and utilized an excellent defensive position. The Chew House, or Cliveden, was a large stone manor belonging to Judge Benjamin Chew. Like the cascines encountered by the French and Austrians in 1734, the British immediately seized on the importance of this potential fortress. Martin Hunter recalled, 'Colonel [Thomas] Musgrave... fortunately now threw himself with the 40th Regiment into [Chew's] House, a very large fine building between Germantown and Beggarstown.'⁴⁰ General Sullivan recalled the situation with concern: 'the enemy had thrown a large body of troops into Chew's House... this was ...very difficult as the house being stone was almost impenetrable by cannon and sufficient proof against musketry.'⁴¹ The Continental front line, however, had passed by the Chew House before the threat became fully apparent. Washington's aide Timothy Pickering believed:

---

34   Hammond, *Letters and Papers of Major-General John Sullivan*, vol.1, p.544–545.
35   Hammond, *Letters and Papers of Major-General John Sullivan*, vol.1, p.576.
36   Martin, *A Narrative of some Adventures*, p.53.
37   Hamilton B. Tompkins (contributor), 'Contemporary Account of the Battle of Germantown', *Pennsylvania Magazine of History and Biography*, vol.11 (1887), pp.330–331.
38   Martin, *A Narrative of some Adventures*, p.53.
39   Philander D. Chase and Edward G. Lengel, *The Papers of George Washington: Revolutionary War Series* (Charlottesville: University Press of Virginia, 2001), pp.401–402.
40   Hunter, *The Journal of General Sir Martin Hunter*, p.34.
41   Hammond, *Letters and Papers of Major-General John Sullivan*, vol.1, p.545.

Sullivan, with his column, had passed Chew's house without annoyance from it... it must have taken some time for Colonel Musgrave, who entered it with six companies of the fortieth regiment, to barricade and secure the doors and the windows of the lower story. Before he would be ready to fire from the chamber windows; it was from them that I the firing I saw proceeding.[42]

Thus, the American force was now in a situation where their front line was engaged with British troops near Germantown proper, while the reserve corps was delayed by the enemy threat at the Chew House. This presented Washington with a conundrum. He wrote to John Hancock after the battle: 'having previously thrown a party into Mr. Chew's House, who were in a situation not to be easily forced and had it in their power from the windows to give us no small annoyance, and in a great measure obstruct our advance.'[43] At this point, there was a famous debate between the officers surrounding Washington over what to do regarding the Chew House, General Henry Knox's party eventually winning out with a decision to focus on the house as a threat rather than leave a small blocking party.[44]

Before, during, and after Washington's preoccupation at the Chew House, another major crisis was unfolding for the Continentals at Germantown: the troops were beginning to run low on ammunition. Even before taking fire from the Chew House, Washington had sent Timothy Pickering forward to General Sullivan. Pickering recalled hearing, 'in advance of us... a very heavy fire of musketry... This fire, brisk and heavy, continuing, General Washington said to me, "I am afraid General Sullivan is throwing away his ammunition; ride forward and tell him to preserve it."'[45] This description of a brisk and heavy fire, combined with Washington's fears of Sullivan 'throwing away' his ammunition, implies a breakdown of fire into a *plackerfeuer* style rolling firefight. Despite Washington's repeated orders to avoid firing at too great a distance, Pickering admitted that he was slightly confused by Washington's order:

I do not know what was the precise idea, which at that moment struck the mind of the General. I can only conjecture, that he was apprehensive that Sullivan, after meeting the enemy in his front, kept us his brisk and incessant fire, when the haziness of the air... prevented his troops from having such distinct view of the enemy, as would render their fire efficient.[46]

Sullivan noted that his men, 'with scarcely a cartridge left, [had] in a severe fire of three hours, expended the whole'.[47] A multitude of sources imply that by this point in the fight, the 40 rounds that the main body of Sullivan's troops had gone into action

---

42 Pickering, 'Battle of Germantown', *The North American Review*, vol.23 (1826), p.427.
43 LOC: Washington Papers, George Washington to John Hancock, 5 October 1777.
44 Pickering, 'Battle of Germantown', *The North American Review*, vol.23 (1826), p.428.
45 Pickering, 'Battle of Germantown', *The North American Review*, vol.23 (1826), p.429.
46 Pickering, 'Battle of Germantown', *The North American Review*, vol.23 (1826), p.427.
47 Hammond, *Letters and Papers of Major-General John Sullivan*, vol.1, p.576.

with was expended. At the Battle of Germantown, Lieutenant Colonel Adam Hubley reported that almost every unit but his own 10th Pennsylvania had 'expended forty rounds' after a firefight that lasted 'four hours, without the least intermission.'[48] As the Continental contingent ran short on ammunition, the unique challenges of the British forces in the Chew House became apparent. Sullivan describes the confluence of factors that caused his men to retreat: 'their cartridges all expended… alarmed by the firing at the Chew House so far in their rear, and by the cry… on the right that the enemy had got round us [they] retired with as much precipitation as they had before advanced.'[49]

Joseph Plumb Martin also recalled this crisis point in the battle:

> Affairs went well for some time. The enemy were retreating before us, until the first division that was engaged had expended their ammunition. Some of the men unadvisedly calling out that their ammunition was spent, the enemy were so near that they overheard them, when they first made a stand and then returned upon our people, who, for their want of ammunition and reinforcements, were obliged in their turn to retreat, which ultimately resulted in the rout of the whole army.[50]

Continental officers wrote recriminating letters years after the battle, carefully challenging any suggestion that they had more ammunition. When it was implied that the Continentals had 60 rounds going into the fight, Timothy Pickering finished a four-page letter to John E. Howard with the questions: '…where or how did the soldiers carry the sixty rounds of cartridge with which Pincheney says each man was supplied? Was your regiment so supplied?'[51] Pickering closed his narrative of the fight at Germantown by saying, 'I had remained near [Washington] until our troops were retreating; when I rode off to the right to endeavor to stop and rally those I met retiring… but it was impracticable; their ammunition I suppose, had been generally expended.'[52] As the troops streamed by Washington, the men were 'holding up their empty cartridge boxes to show him why they ran.'[53]

Washington's forces drew off from Germantown, but the army was less concerned by the defeat. Two North Carolina officers reported to their state: 'the spirits of our Soldiers are great in consequence of their having discovered that they can make their enemies run.'[54] The radical Thomas Paine, who accompanied the American army to the battlefield noted, 'the retreat was extraordinary. Nobody hurried themselves.

---

48 Quoted in Thomas McGuire, *The Philadelphia Campaign, Volume II: Germantown and the Roads to Valley Forge* (Mechanicsburg: Stackpole, 2007), p.103.
49 Hammond, *Letters and Papers of Major-General John Sullivan*, vol.1, p.546.
50 Martin, *A Narrative of some Adventures*, p.54.
51 Historical Society of Maryland, Bayard Papers, Timothy Pickering to John E. Howard, 10 February 1827.
52 Pickering, 'Battle of Germantown', *The North American Review*, vol.23 (1826), p.429.
53 Christopher L. Ward, *The Delaware Continentals, 1776–1783* (Wilmington: The Historical Society of Delaware, 1941), p.229.
54 North Carolina Department of Archives and History: Executive Letter Book, John Penn and Cornelius Harnett to Richard Caswell, 20 October 1777, p.659.

Every one marched his own pace... They appeared to me to be only sensible of a disappointment, not a defeat'.[55] In any case, the Americans had suffered heavier losses. The American total of killed, wounded, missing, and captured was almost 1,100; the British total was slightly over 500.[56]

The fighting at Germantown also illustrates many of the themes of this book. Officers like Washington placed a high degree of faith in the ability of troops with charged bayonets to chase their enemies from the field, preferably without firing. While the bayonet was a powerful psychological tool, American forces clearly utilized a great deal of firepower, as evidenced by the heavy ammunition expenditure during the four-hour battle. As was so often the case, hopes that the bayonet would prove decisive faded once the battle had degenerated into a firefight, the sort of engagement that Washington feared, and tried to use his most precious instrument of battlefield control (his aides) to prevent. A heavy firefight eventually led to American forces using up the 40 rounds that they had brought to the battlefield and withdrawing. For their part, the British used a style of defensive fighting that would have seemed perfectly at home on the battlefields of Parma or Guastalla: they carefully selected fighting positions which covered their men from enemy fire, utilizing every wall, hedge, ditch, fence, and yard on the battlefield. To some extent, this can help explain the lower casualties on the British side. Most importantly, they seized a key piece of terrain: a large stone house running along the American axis of advance. The Chew house filled the same role at Germantown that the cascines had played at Parma and Guastalla.[57]

Eutaw Springs, 8 September 1781

Operationally, American General Nathanael Greene found himself in a very similar position to Washington at Germantown in 1781. With a larger force, Greene felt pressure to assume the offensive in order to drive the British forces in South Carolina back towards Charleston, before hopefully liberating that city. Otho Holland Williams, a Maryland officer in Greene's army, wrote to his friend that Greene had no greater reason to fight at Eutaw Springs: 'Greene did not approve of their holding that post, and as his forces were now collected, he determined to prosecute his plan of giving battle or removing them to a peaceful distance.'[58] With the decision to fight taken, Greene brought his approximately 2,000 men to bear on the roughly 1,400 men of Lieutenant Colonel Alexander Stewart.[59]

---

55 William B. Wilcox, *The Papers of Benjamin Franklin* (New Haven: Yale University Press, 1987), vol.26, pp.479–489.
56 McGuire, *The Philadelphia Campaign, Volume II*, p.128.
57 In an odd twist of fate, Cliveden was painted as a striking Italian manor house in the Neapolitan artist Saviero (Xavier) della Gatta's famous depiction of the Battle of Germantown. See the front cover of the book.
58 Historical Society of Maryland: Otho Holland Williams Papers, MS 908 Item 116, Otho Holland Williams to Major Edward Giles, 23 September 1781.
59 Robert Dunkerly and Irene Boland, *Eutaw Springs: The Final Battle of the American Revolution's Southern Campaign* (Columbia: University of South Carolina Press, 2017), p.31.

Despite being more willing to seek battle late in the war, Nathanael Greene did not emulate Washington and Lacy at Eutaw Springs. Rather, in this battle, Greene marched to the field and deployed in a standard series of lines. He did, however, attempt to emulate the success of Daniel Morgan at Cowpens, reconfiguring the deployment of that general for an offensive battle. Thus, at Eutaw Springs, Greene ordered the militia to the attack first, followed by relatively inexperienced Continental troops, capped off by an attack with his third line of Continental veterans.[60] This sequence had the happy consequence of breaking through the British battle line, before ammunition shortages and enemy fallback defences forced the Continentals to withdraw.

Lieutenant Colonel Stewart led a force that was badly ailing in want of provisions. To begin with, he was camped at Eutaw Springs, 'where I might have the opportunity of receiving my supplies and disencumber myself form the sick.'[61] Desperately short of provisions, Stewart had taken to sending out 300 men in the mornings to dig for sweet potatoes. The Otho Williams reported, 'an unarmed party, under a small escort, had been advanced up the river for the purpose of collecting the sweet potato… to contribute to the subsistence of his army.'[62]

The battle was fought on an east-west axis, with the battlefield bounded to the north by the Santee River and Eutaw Creek. The Americans advanced along a road which ran east to west. The British were camped in an open potato field near a large brick house, which had a palisaded garden to its north along the banks of the river. The house and garden complex also included a number of outbuildings. Further to the west of the house, garden, and field, lightly wooded country ran along the south bank of the Santee River. Otho Williams described the terrain:

> The whole country on both sides of the road being in woods… the woods were not thick, nor the face of the country irregular… bounded north by the creek… and has a high bank thickly bordered with brush and low wood. From this house to the bank, extended a garden enclosed… the house was brick, and abundantly strong to resist small arms.[63]

On the morning of 8 September, Stewart had sent out his rooting party as normal, and then received intelligence that Greene was advancing on his position. As a result, he dispatched Loyalist Captain John Coffin with a party of around 200 men, with orders to discern Greene's intentions and recall the rooting party. Tragically for Coffin and Stewart, the Continentals fell in with the scouting party, the rooting

---

60 Nathanael Greene and Dennis M. Conrad, *The Papers of General Nathanael Greene* (Chapel Hill: University of North Carolina Press, 1997), vol.9, p.329.
61 'Extract of a letter from lieut. Col. Stewart to Earl Cornwallis', in Robert Gibbes, *Documentary History of the American Revolution: consisting of letters and papers relating to the contest for liberty* (New York: Appleton, 1857), vol.3, p.137.
62 'Account furnished by Col. Otho Williams', in Gibbes, *Documentary History of the American Revolution*, vol.3, p.145.
63 'Account furnished by Col. Otho Williams', in Gibbes, *Documentary History of the American Revolution*, vol.3, p.147.

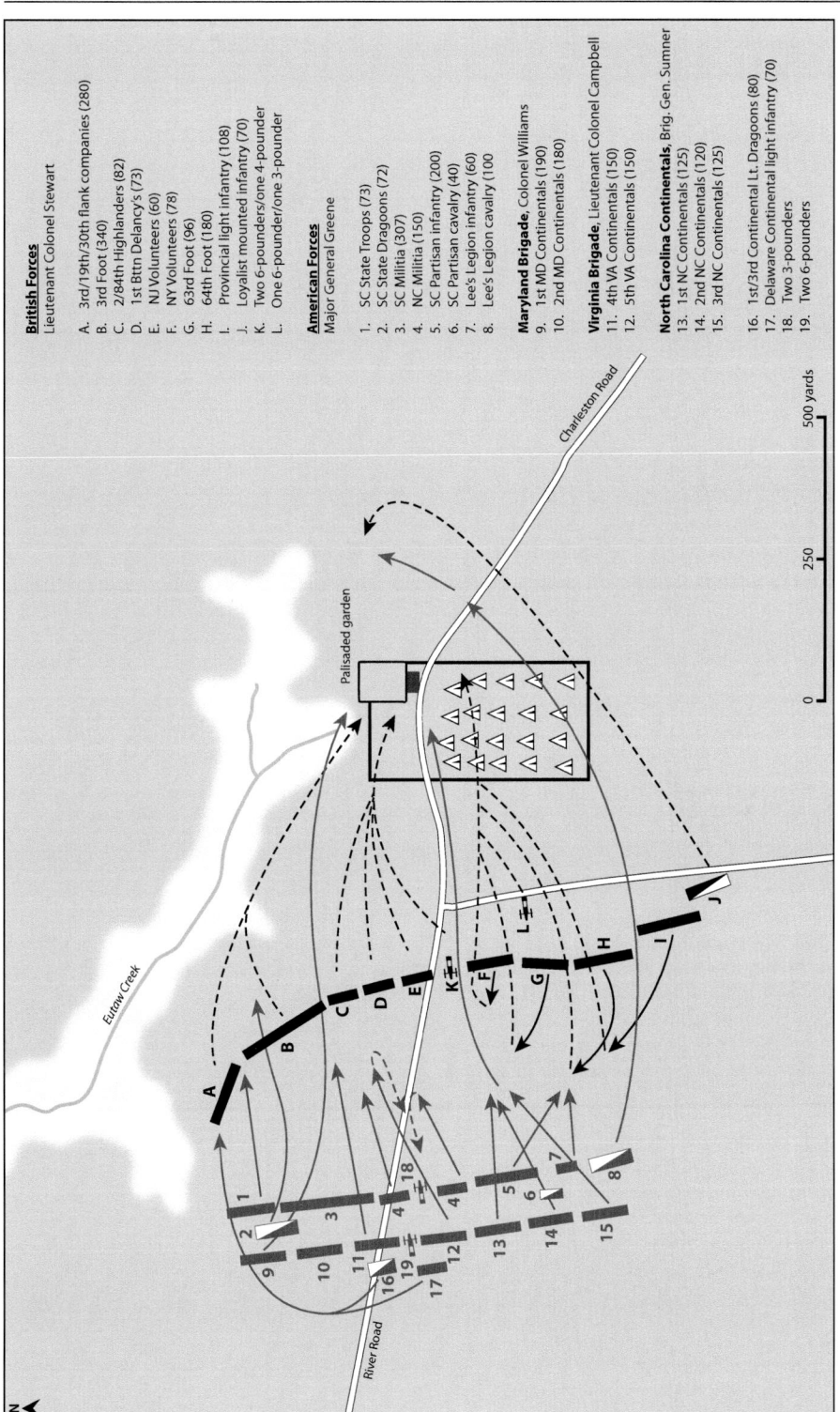

The Battle of Eutaw Springs, 8 September 1781.

party came to investigate the disturbance, and several hundred men were put out of action or captured by the Americans.[64]

At this point, Stewart began to realize the scale of the problem with which he was confronted, and deployed skirmishers to delay Greene's advance while he formed the rest of his command in order of battle. Williams noted, 'Stewart had pushed forward a detachment of infantry to a mile distant from the Eutaws, with orders to engage and detain the American troops.'[65] Major General Greene noted, 'The enemy's advanced parties were soon driven in.'[66] Otho Williams, who had used a similar tactic against the British at the Battle of Camden in 1780, noted, 'the skirmishing parties were cleared away from between the two armies.'[67] In a letter to Major Edward Giles written shortly after the battle, William's was more direct: 'the enemy's van was soon driven to their line and our troops displayed.'[68] Here, Williams uses the typical American language for forming the line of battle ('displaying'), indicating contrary to his later statements on the battle, that the battle line did not form the line of battle immediately after engaging Coffin.[69] In this version of events, at least, Williams argues that the American troops remained in a column formation until the enemy skirmishers had been driven in.

At this point, it was up to the American first line of militiamen from North and South Carolina to try and drive back the British regulars. Otho Holland Williams asserts that they fired 17 rounds before breaking away from the British troops. The militiamen themselves claimed that 'they fired fifteen to twenty rounds each man' and 'on the fire of twelve rounds we obliqued to the right, still firing.'[70] This was truly an impressive performance for the militia, especially considering that the army had drawn ammunition according to Greene's order from 30 August: 'Each Continental soldier will receive thirty cartridges, each militiaman twenty.'[71] Thus, before withdrawing, the militia had used almost all of their ammunition supply. It was now

---

64 Dunkerly and Boland, *Eutaw Springs*, p.36.
65 'Account furnished by Col. Otho Williams', in Gibbes, *Documentary History of the American Revolution*, vol.3, p.146.
66 Nathanael Greene and Dennis M. Conrad (ed.), *The Papers of General Nathanael Greene* (Chapel Hill: University of North Carolina Press, 1997), vol.9, p.329.
67 'Account furnished by Col. Otho Williams' in Gibbes', *Documentary History of the American Revolution*, vol.3, p.146. See also Historical Society of Maryland: Otho Holland Williams Papers, MS 908 Item 115, Otho Holland William to Elie Williams, 11 September 1781.
68 Historical Society of Maryland: Otho Holland Williams Papers, MS 908 Item 116, Otho Holland Williams to Major Edward Giles, 23 September 1781.
69 'Account furnished by Col. Otho Williams', in Gibbes, *Documentary History of the American Revolution*, vol.3, p.146. For American soldiers using the term 'displaying' to describe forming the line of battle, see: Jeremiah Greenman, *Diary of a Common Soldier in the American Revolution* (DeKalb: Northern Illinois University Press, 1978), p.173; Society of the Cincinnati Library, MSS L2007G37, Diary by an officer of the Third Pennsylvania Continental Line, May 26, 1781-July 4, 1782; William Feltman, *Journal of William Feltman of the First Pennsylvania Regiment*, (Philadelphia: Historical Society of Pennsylvania, 1853), p.7; NAM: 1991-09-117, Captain John Davis; LOC: Washington Papers, William Heath to George Washington 9 November 1781.
70 Quoted in Patrick O'Kelley, *Unwaried Patience and Fortitude: Francis Marion's Orderly Book* (West Conshohocken: Infinity, 2006), p.547.
71 Greene and Conrad, *The Papers of General Nathanael Greene*, vol.9, p.271.

left to the better-equipped Continental troops to try and drive the British forces back from their line just inside the woods on the edge of the Eutaw potato field. The battle lines engaged at close range, facilitated by the cover of the woods, which may have even led to a higher-than-normal incidence of hand-to-hand combat.[72] In the end, despite weakening the British forces, the militia withdrew from close combat, leaving the American Continentals to finish the fight.

North Carolina Continental troops under General Jethro Sumner now moved forward to replace the militia. Most sources are silent regarding this part of the battle. In his correspondence directly after the battle, Williams writes that the North Carolinians 'regained honor by their firmness,' and Greene notes that, 'though not above three months raised, [they] behaved nobly.'[73] Reading between the lines, it seems as though the North Carolinian Continentals were badly outnumbered: 350 of them faced the entire British line and withdrew. Depending on the perspective of the source, this either was caused by, or caused itself, the left of the British line to advance to the charge. Stewart asserts that this attack drove off both the militia and the North Carolina troops, while Williams argues that it was the retreat of the North Carolina troops that tempted the British to advance.[74]

At this juncture, widely celebrated by American accounts of the battle, the combined Virginian and Maryland troops, approximately 600 men, launching a bayonet attack, 'advanced in good order with trailed arms and without regarding or returning the enemy's fire charged and broke their best troops.'[75] Perhaps unsurprisingly, Stewart disputes this good behaviour on the part of the Americans, saying that the Virginians and Marylanders broke his line with 'a heavy fire.'[76] Regardless of the cause of the rout, the British main line fell back in haste to the brick house, walled garden, and heavy thickets on the bank of Eutaw Creek, pursued by the victorious Americans.

The American cavalry under Colonel William Washington and Lieutenant Colonel Henry 'Light Horse Harry' Lee then attacked the British right and left flanks, respectively. Although Lee was successful on the southern end of the battlefield against the British left, on their right flank, Washington:

> Impatient perhaps of a more favorable opportunity, charged upon the enemy's right, where unluckily their flank companies were posted. He received a very galling fire by which his horse fell in the front of his dragoons. In an instant his breast was pierced with a bayonet, which wounded him but slightly. His cavalry was repulsed, and that excellent officer [was captured].[77]

---

72   Dunkerly and Boland, *Eutaw Springs*, p.49.
73   Historical Society of Maryland: Otho Holland Williams Papers, MS 908 Item 116, Otho Holland Williams to Major Edward Giles, 23 September 1781; 'General Greene to the President of Congress', in Gibbes, *Documentary History of the American Revolution*, vol.3, p.143.
74   Gibbes, *Documentary History of the American Revolution*, pp.137, 149.
75   Historical Society of Maryland: Otho Holland Williams Papers, MS 908 Item 116, Otho Holland Williams to Major Edward Giles, 23 September 1781.
76   'Extract of a letter from lieut. Col. Stewart to Earl Cornwallis', in Gibbes, *Documentary History of the American Revolution*, vol.3, p.137.
77   Historical Society of Maryland: Otho Holland Williams Papers, MS 908 Item 116, Otho Holland Williams to Major Edward Giles, 23 September 1781.

At this point, the main American effort was directly against the enemy infantry posted in heavy cover from the thickets to the garden and to the brick house. American troops chased the British through their camp, where, as Robert Dunkerly and Irene Boland have argued, looting of the camp has probably been overstated as a reason for American defeat, since it does not appear in many firsthand accounts.[78] Instead, American troops pursuing the British into their camp found themselves in desperate need of cover, as Otho Williams argued: 'Nor was the concealment afforded by the tents at this time a trivial consideration, for the fire from the windows of the house was galling and destructive, and no cover from it was anywhere to be found expect among the tents, or behind the buildings to the left of the front of the house.'[79] This was increasingly becoming a desperate situation for the American attackers, as they lost momentum and more and more men began to seek cover wherever they could in the face of well-protected British firepower from multiple angles. The American officers tried to bring their men through the camp on the other side, but in doing so exposed themselves. Several key American infantry commanders, including John Eager Howard, were wounded in the space of a few moments.[80] Williams noted, 'when their officers had proceeded beyond the encampment, they found themselves nearly abandoned by their soldiers, and the sole marks for the party who now poured their fire from the windows of the house.'[81] Just after the battle, he was more forthright: 'Their fire began to gall us exceedingly. About this time, Major General Greene had brought our two six pounders within about one hundred yards of the house, and by accident or by mistake, two others which we had taken were brought to the same place.'[82]

Greene hoped that by using artillery, he would be able to engage the enemy targets more successfully in cover. In his own battle report, Greene described the situation:

> Four cannon were advanced against the house, but the fire from it was so brisk, that it was impossible to force it, or even to bring o[ff] the cannon, when the troops were ordered to retreat, and the greatest part of the officers and men who served those cannon were either killed or wounded… seeing our Foot roughly handled by the enemy's fire, and our ammunition almost expended, I thought it my duty to shelter them from the fire of the house.[83]

This action demonstrates the heavy advantages enjoyed by troops firing from cover. Greene had the artillery drawn into a range of 100 yards, what most secondary works argue is at the limit of accurate range for muskets. At this distance, the

---

78 Dunkerly and Boland, *Eutaw Springs*, pp.65, 70.
79 'Account furnished by Col. Otho Williams', in Gibbes, *Documentary History of the American Revolution*, vol.3, p.154.
80 Dunkerly and Boland, *Eutaw Springs*, p.65.
81 'Account furnished by Col. Otho Williams', in Gibbes, *Documentary History of the American Revolution*, vol.3, p.154.
82 Historical Society of Maryland: Otho Holland Williams Papers, MS 908 Item 116, Otho Holland Williams to Major Edward Giles, 23 September 1781.
83 'General Greene to the President of Congress', in Gibbes, *Documentary History of the American Revolution*, vol.3, p.143.

British troops firing from the cover of the house were not only able to pick off enemy officers, but also kill and wound the artillerymen and their officers working the guns. As a result, Greene not only lost his two six-pounders, but also those taken from the British when their first line collapsed. This was the turning point of the battle, as Williams asserted:

> At this critical juncture the enemy made a conclusive effort which not only did them great honor, but in my opinion was the salvation of their whole army. Major [John] Majoribanks sallied briskly from behind [the] picketed garden, charged our artillery, and carried the pieces, which they immediately secured under the walls of their citadel… As our two three-pounders and one which we had taken in the field were all dismounted it was useless to attempt any thing further with small arms. The general therefore ordered the troops to retire, which was done gradually.[84]

With the British forces so greatly weakened by the losses of the battle and the detached parties in the opening phase of the engagement, Lieutenant Colonel Stewart elected to let Major General Greene retire from the field without pursuit. The British, losing a large number of prisoners at the opening of the engagement, lost between 700–900 men. The Americans likely suffered a greater number of killed or wounded, but fewer prisoners, for a total of around 500–600 men.[85] Greene claimed, much like the Americans at Germantown, that he had withdrawn due to a lack of ammunition.[86] Logistical reports from his army support this idea: over a month after the battle on 10 October, most of the Continental troops could still only be issued around 10 rounds per man.[87] Since the militia shot away most of their ammunition, the situation was likely even worse for them, but no returns of their ammunition survive. A month later in November, Greene wrote a frustrated note, saying that his troops 'should have marched eight or ten days since but for the want of ammunition not having ten rounds a man.'[88] Ammunition proved a severe headache, even for troops who prided themselves on the ability to 'sweep the field with [their] bayonets.'[89]

In many ways, Eutaw Springs was a miniature Germantown: the American forces, using a novel style of grand tactics pioneered by Daniel Morgan at Cowpens, had inflicted heavy losses on the British. However, they were unable to drive the redcoats from the field as a result of British use of tactical terrain, specifically the brick house. A shortage of ammunition plagued the Americans as much as the British fire, and fire proved accurate against individual officers and artillerymen, even out to 100

---

84 'Otho Holland Williams to Major Edward Giles', September 23rd, 1781, MS 908 Item 116, Otho Holland Williams Papers, Historical Society of Maryland
85 Dunkerly and Boland, *Eutaw Springs*, p.86.
86 'General Greene to the President of Congress' in Gibbes, *Documentary History of the American Revolution*, Volume 3, p.143.
87 'Returns of Ammunition October 1781' MS 908, Items 117–121, Otho Holland Williams Papers, Historical Society of Maryland.
88 Greene and Conrad, *The Papers of General Nathanael Greene*, vol.9, p.598.
89 'Account furnished by Col. Otho Williams' in Gibbes, *Documentary History of the American Revolution*, vol.3, p.149.

yards. In many ways, Eutaw Springs is a reminder of the flexibility of British forces when on the tactical defensive: engaged against a numerically superior opponent, Lieutenant Colonel Alexander Stewart delayed the enemy with skirmishers and survived the collapse of his main line by withdrawing to cover, where he could wear down Greene's superior forces.

Conclusion

Despite the 50 years which separate them, many similarities connect Parma, Guastalla, Germantown, and Eutaw Springs. In a world where officers prized troops that manoeuvred with precision and attacked with their bayonets without firing, the ability of troops to carefully aim their weapons and utilize skirmishers to delay the enemy marked all of these battles. Troops fired on the enemy without orders and were sheepish about this in reports after the battle. They fired at longer ranges than many twentieth-century historians admit, and frequently fired independently of their officers' commands. At Parma, Guastalla, Germantown, and Eutaw Springs, ideas about approaching the enemy closely with loaded muskets or bayonets occasionally worked to intimidate them, but more frequently the side which chose defensive terrain which facilitated accurate firing from cover was victorious. The cascines of the Po River Valley in the 1730s have their counterparts in Cliveden at Germantown and the brick house at Eutaw Springs.

At all four battles, cavalry failed to play a truly decisive role. These battles provide a good case study, then, of what infantry warfare was like across military Europe when cavalry was not able to make its presence felt on the battlefield. This was not always the case: in a European battle, cavalry could be highly decisive. These battles are also connected in that they represent a force normally associated with defensive tactics taking the offensive against a more tactically aggressive opponent. More often than not, in these four battles, the side which embraced the realities of eighteenth-century infantry combat, rather than the side which attempted to use preconceived notions regarding intimidating bayonet attacks, emerged victorious. The current author leaves it to the reader whether or not these four battles constitute enough of a sample from which to draw a pattern onto eighteenth-century warfare.

10

Conclusion

Infantry in Battle, 1733–1783

You awake feeling stiff, as a light rain soaked where you had lain overnight.[1] Tents and blankets had remained with the baggage train with the enemy so near, so your cow-hide bag was your pillow, the earth was your bed, and the sky was your blanket.[2] It is not yet light, and your battalion assembles to march before the sun rises. You move out, marching in a closed column of platoons. You have known many of the men marching in your platoon for years: included amongst you are your uncle and village neighbours.[3] You move roughly along a road.

As the sun rises, your march continues. You begin to hear pops in the distance, followed by silence, and then what could be the distant peal of thunder. Eventually, to your ears, the noise builds to something that sounds like the burning of new thorn branches under a kettle and a constant thunderstorm.[4] The men of your platoon, and indeed the officers, begin to look a bit concerned.[5] A group of mounted officers rides passed to gloomily confer with your battalion commander. The men begin to murmur; it is the king! He pauses briefly near your platoon, yelling in a thin reedy voice, 'Good Morning, lads!' Your platoon returns the greeting, and as the king rides off, he turns in the saddle and shouts back: 'Well, boys, do you want to come along?' The battalion dissolves into cheers of 'yes, yes!'[6]

Your battalion marches down a wooded lane, following the next battalion in your brigade. You emerge from the woods into a more open farming country, and the thunderous roar of the artillery becomes more urgent. You hear shouted commands from your officers, and your battalion, and those of your brigade, begins the urgent

---

1 Zander, *Fundstücke*, p.62.
2 This is a direct quotation from: 'Briefe Preußischer Soldaten aus den Feldzügen 1756 und 1757', in Curt Jany, *Urkundliche Beiträge Und Forschungen Zur Geschichte Des Preussischen Heeres* (Berlin: E.S. Mittler, 1901), p.2.
3 Zander, *Fundstücke*, p.33.
4 Hamilton B. Tompkins, 'Contemporary Account of the Battle of Germantown', in *Pennsylvania Magazine of History and Biography* (1887), vol.11, pp.330–331.
5 Ulrich Bräker and Johann Heinrich Füssli, *Lebensgeschichte und natürliche Ebentheuer des Armen Mannes im Tockenburg* (Zürich: Füssli, 1789), p.148.
6 Anon., *Offizier-Lesebuch, historisch-militärischen Inhalts, mit untermischten interessanten Anekdoten, von einer Gesellschafts Militärischer Freunde* (Berlin: C. Matzdorff's Buchhandlung, 1793), pp.183.

process of deploying into line. Each of the platoons march at a quicker step towards its appointed place, and the line is formed. The enemy artillery is now targeting your battalion directly. Great clouds of earth shower the line with near misses.

About 1,000 yards away, through the walled fields of grain and tree-lined paths to your front, there is a sudden puff of smoke. The thunderclap of a cannon follows two second later, and a round shot passes very close to the head of a man down the line in your platoon. He jerks and falls, blood flowing from his nose and mouth.[7] The line advances at a steady pace, moving through fields of grain. More clouds of earth as round shot skips over the unit. Then, a file away, a round shot tears into a man from your village that you have known for years. His blood, fragments of bone, and brains spray your face.[8] The line buckles and reforms as you close the gap left by his file. The officers are shouting, 'Lads, behave like the brave soldiers you are. We won't abandon you, and don't you abandon us!'[9]

You pass out of the grain field, crossing a low wall and into a tree-lined path. The officers order a halt, and you crouch in the slight cover afforded by the sunken path. The battalion guns that have accompanied your battalion this far now fire at the distant enemy position. Then you see the reason for the halt. Off to your right, a small village separates you from the next battalion. The villagers have dammed a stream for a fishing pond directly across your line of march, and the officers are discussing how to best cross it. The battalion to your left, unimpeded, continues its advance ahead of you. There is now a gap in the battleline. The enemy artillery has moved to targeting that battalion, since yours is obscured. Your battalion commander rides up, dismounting from his horse. With a shout of 'Lads, follow me!' he starts running towards the pond. Not strictly keeping in their ranks and files, the battalion runs after him.[10]

Wading across the pond or crossing at a small walkway, the battalion struggles over the obstacle. The battalion is no longer in order, soldiers still straggling over the pond. The battalion commander and his captains gather the colour party, form them facing the enemy, and order the drummers to beat *Alarme*. At this signal, the men race to reform the battalion facing the enemy. In a few moments, this is achieved, and the battalion marches forward, more or less aligned with its fellows to the right and left.[11]

The enemy artillery fire has switched to cannister, but mercifully they are overshooting the battalion, which seems to be advancing quicker than they expected. Then, a noise sounds like a rush of wind, and a clattering as dozens of canister shots hit the upright bayonets of your marching platoon. A lieutenant's spontoon is torn

---

7   Joseph Plumb Martin, *A Narrative of some Adventures, Dangers, and Sufferings of a Revolutionary Soldier* (Hallowell: Glazier, Masters & Co, 1830), p.173.
8   Curt Jany, 'Briefe Preußischer Soldaten aus den Feldzügen 1756 und 1757', p.30.
9   Quoted in Duffy, *The Army of Frederick the Great* (2020), p.103.
10 Anon., *Sammlung ungedruckter Nachrichten so die Geschichte der Feldzüge der Preußen von 1740 bis 1779*, (Dresden: Walther, 1785), vol.4, p.118; Friedrich Christoph Fischer, *Geschichte Friedrichs des Zweiten: Königs von Preussen* (Halle: Francke, 1787), vol.1, p.504.
11 *Reglement vor die Königlich Preussische Infanterie* (Berlin, Given and Printed June 1st, 1750), pp.131–132.

from his grasp as a canister ball hits the blade.[12] There are cries, sickening thuds, and groans, as the cannister finds its mark in the next platoon.

Three hundred yards away, waiting for you along a low stone wall, you see an enemy infantry battalion. You approach, slowly, keeping step, to within 200 yards. They fire together as a battalion, smoke obscuring their position for a moment. You listen, but hear few cries from wounded men, at least not in your platoon. Your officers are shouting for the men to continue advancing on the enemy. You hear an almost pleading note as they repeatedly order the men to leave their muskets at the shoulder.[13] The advance is steady. The enemy is 150 yards away. The enemy fires again. One hundred yards. Somewhere along the battalion, perhaps two platoons away, someone fires.[14] You immediately tear your musket from your shoulder and fire.

Chaos. The entire battalion discharges their weapons. Officers are screaming at the men to cease fire, hitting the men with canes and the flats of their swords.[15] Your battalion commander rides up and down the line, ordering the men to shoulder their arms. The enemy has fired again; this time men are hit. Your platoon reloads and shoulders their muskets, others do the same. The order is given to advance, and hesitatingly, not perfectly in unison, the advance continues. Emerging through the smoke of your own musketry, you clearly see the enemy battalion, barely 75 yards away. The enemy fires again. More men are hit. The man on your right is killed, and someone fills his place from the rear. The battalion commander orders, 'March, March!' Your platoon breaks into a jog, muskets held across the chest.[16] The enemy begins to flee.

You cross the former position of the enemy platoon, where the enemy lie wounded, moaning, and dying on the ground. Some are finished off by men from your battalion.[17] An officer asks if you are wounded and you immediately reply 'no' before realizing that your coat has several holes, one of which is bloody.[18] The only hole with blood is in your side, from a very small-calibre ball, essentially a buckshot. It was fired at too great a distance, bruising you and breaking the skin, but not, you hope, doing much damage.[19] Now, forming the battalion, your officers order the battalion to wheel to assault the artillery battery that has been tormenting you since the beginning of the engagement. The guns are barely 300 yards away.

---

12   Carl Wilhelm von Hülsen, *Unter Friedrich dem Großen: aus den Memoiren des Aeltervaters* (Berlin: Paetel, 1890), pp.88–89.
13   Wilson, *The Life and Letters of James Wolfe*, p.37.
14   Anon., *Sammlung ungedruckter Nachrichten*, vol.1, p.359; Historical Manuscripts Commission, *Report on Manuscripts in Various Collections*, vol.8, p.408.
15   Historical Manuscripts Commission, *Report on Manuscripts in Various Collections*, vol.8, p.408.
16   Anon., *Schreiben eines Holländischen Volontairs* (Dresden: Publisher Unknown, 1757), vol.9, p.5.
17   Bräker and Füssli, *Lebensgeschichte*, p.154.
18   Nathaniel Root, 'The Battle of Princeton', *Pennsylvania Magazine of History and Biography*, vol.20 (1896), p.518.
19   Anon., *Besondere Merkwürdigkeiten und Anekdoten aus Neudam in der Neumark* (Salzwedel: J.C. Schuster, 1758), p.29.

You begin the march up the hill. At 200 yards away, the guns of the battery discharge, one after another, seemingly aimed at the colour party in the midst of the battalion. The colour party and the platoons on either side of them seem to almost disintegrate in a hail of lead. A number of officers are killed. The advance falters. The men begin to fire at the guns, but at this distance and against a relatively small target, will it do much good? Individual men, particularly near the centre of the battalion, begin to run in ones and twos back down the ridgeline towards the stone wall you started from. Junior officers curse the men, striking out with the flats of their swords. Your battalion commander, though, is heard shouting over the sound of the din: 'Let them run! We will bring them together over there!' He points back towards the stone wall with his cane. The unit breaks, and in a short time, is reformed by the stone wall. In the intervening span of time, the next battalion on your right has assaulted the battery themselves, taking the guns. Shaken, the battalion gathers itself together.[20]

Reforming, you continue to advance up the hill, through a wood at the top of the ridgeline. As soon as you have gotten clear of the trees, you see infantry from of the enemy's second line advancing on you, barely 150 yards away. The officers shout, 'Platoon: Ready! Present! Fire!' and the front of the platoon is obscured by smoke and the roar of the musketry.[21] The enemy returns fire, but your men load more speedily. They also suffered the losses of your first volley, so it is an even fight. The commander of your brigade, riding over from the battalion on your right, trots up and down the line behind your battalion in thick small-arms fire, shouting to the men, 'Lads, hold fast! Fire bravely! The enemy will soon run!'[22] The battalion guns and supporting artillery have caught up with the advance of the infantry at long last. They join the fight, firing canister at the enemy infantry. Your battalion commander offers the gunners a monetary reward if they can drive off an enemy battery.[23]

The enemy begins to waver, and driven on by the shouts of the officers, your platoon starts to bound towards them, passing from wall to wall, and hedge to hedge on your way down the ridge, firing from cover as you go. The battalion takes up the shouts of 'Victory!' and is no longer in perfect order but moving pell-mell through the agricultural landscape. You pass individual and knots of wounded men. You shout, 'Come on, brothers!' Some join you; others continue to make their own way backwards or to the side.[24]

As you move down the ridge, on a plain in the distance you see the swirling melee of cavalry; whether friend or foe is winning, you cannot tell. Having come down the ridge, you chase the fleeing enemy troops through a small village. Muskets fire from doorways and windows, but the enemy is mostly broken. Your battalion takes positions in the village, and the other battalions of your brigade form to the right

---

20   Georg Heinrich von Berenhorst and Eduard von Bülow, *Auch dem Nachlasse von Georg Heinrich von Berenhorst* (Dessau: Karl Due, 1845), vol.1, p.103.
21   Ernst Friedrich Rudolf von Barsewisch, *Meine Kriegs-Erlebnisse während des Seibenjährigen Krieges, 1757–1763* (Berlin: von Warnsdorff, 1863), p.113.
22   Barsewisch, *Meine Kriegs-Erlebnisse*, p.114.
23   Barsewisch, *Meine Kriegs-Erlebnisse*, p.115.
24   Bräker and Füssli, *Lebensgeschichte*, p.156.

and left of your position. The enemy flees into a wood on the other side of the village, and you do not see them reappear. The advance seems to have stopped. The men from your battalion root through the village, looking for food and drink. The details of wounded men you pass now start to stick in your mind. Some men sit down to write letters home, shouting offers to pass on messages from their comrades in the same village. Many men are ordered to find the wounded from the battlefield and bring them to the surgeons.[25] Within a few hours, the battalion is mostly dispersed, working to aid the wounded, collecting trophies from the battlefield, drinking, eating, or sleeping. As afternoon wears on, you see two hussar officers riding by in the midst of an angry confrontation. One wishes to have the men who have become drunk on village wine beaten for their carelessness; they could have been pursuing the enemy! The other, an older officer, simply says, 'No Colonel, don't do that. Let them enjoy their rest. They have had a hard enough time today.'[26]

(Early) Modern Warfare

This book highlighted the highly negotiated nature of infantry combat in the eighteenth century. Officers and military theorists had their preferred tactics: they wrote extensive treatises on how infantrymen should always hold their fire, be equipped with longer bayonets, have body armour reintroduced, or readopt the pike. Even when not led astray by these fanciful suggestions, historians have often prioritized the theoretical writings of officers on eighteenth-century warfare, rather than the descriptive evidence of what actually occurred. This is an omission which this book corrects. When the writings of enlisted men are utilized frequently alongside the writings of officers, as they have been in this book, a new story begins to take shape, a story like the fictious encounter above. In creating this scenario, 14 of the 26 supporting citations have been drawn from the words of actual infantrymen describing their experiences. Eleven citations have been drawn from the descriptive writings of officers, and one has been drawn from a prescriptive source – a drill manual. Historians of battle should utilize and read all of these sources but give greater priority to the early modern descriptive sources, particularly where those sources are drawn from soldiers rather than officers.

Using this methodology allows for writing a history of battle from below. As this book has shown, officers and men may have different perspectives on the experience of battle, and to understand the human experience of combat more fully, it is necessary to examine both perspectives. This is not a magical solution; it brings with it a host of methodological problems. Just how representative are these letters and memoirs from soldiers? What percentage of these men were really literate? How many of them would have agreed with the perspectives of their literate comrades? All of these points are valid, and must be considered when working with the small number of sources, relatively speaking, that survive from enlisted men in the

---

25 Todd, *The Journal of Corporal Todd*, pp.166–169.
26 J. C. Lojewksy, *Sebstbiographie des Husaren-Obersten von…ky* (Leipzig: Kollmann, 1843), vol.2, p.62.

mid-eighteenth century. Despite these challenges, in order to understand a period of combat, where sources are available, it is vital to hear from both officers and enlisted men.

Of course, there were officers who were broadly sympathetic to the views of enlisted men. In 1738, Alvaro Navia Osorio, Marqués de Santa Cruz de Marcenado, railed against the standard views of many officers presented in this book. 'Many claim,' he argued, 'that one should not fire at the enemy, except when one comes to be scarcely further them than the reach of the bayonet.'[27] Santa Cruz continued, 'I think differently… if you do not start firing at the enemy immediately, you are depriving yourself of the advantage of killing many of them, and intimidating others with the buzz of bullets.'[28] By firing at the enemy, you would benefit from 'the fright this spectacle would cause the recruits and new soldiers of the enemy, who are most perturbed by this… they will take aim with less order, and when you finally approach with edged weapons, they will be much diminished and intimidated.'[29] For Santa Cruz, 'If soldiers only carried enough ammunition to fire a few shots, I would be content to leave this to close range and only fire then. But as the situation is, you must conclude the opposite.'[30] All of these factors combined in Santa Cruz's final rejoinder:

> As soon as the enemy prepares to present, you must have your men kneel down, in order to present a smaller target, so that you will lessen the accuracy of their shots: for, as I have said, soldiers often aim too high. What I have just proposed, it must be said, works best at long range, so the enemy will not have time to approach you closely with the bayonet before your men can get up… near Saint Etienne Litera a detachment of English infantry knelt to the ground when it saw the French in a position to make their fire. Then, it quickly got up, having received no damage.[31]

Many officers understood that an advantage could be gained by fighting in a way which made sense to enlisted men. Simply framing this as a class divide makes little sense, as both officers and men faced the dangers of combat. But what is clear is that soldiers played an important role in driving this process through negotiated authority on the battlefield. They were not simply terrified conscripts hoping that officers would not beat them; they took an active role in obeying, frustrating, and at times, disobeying their officers when they believed the situation required it.

So, who was right: The Prussomaniac officers who were arguing for tighter control, or men like Santa Cruz who were arguing in line with what the men were already doing? The wider century gives three primary examples of armies who set out a tactical doctrine based on bayonet intimidation rather than firepower. The Swedish army of the Great Northern War, the Prussian army of the early Seven

---

27 Santa Cruz, *Reflexions militaires et politiques*), vol.6, p. 66.
28 Santa Cruz, *Reflexions militaires et politiques*, vol.6, p.67.
29 Santa Cruz, *Reflexions militaires et politiques*, vol.6, p.67.
30 Santa Cruz, *Reflexions militaires et politiques*, vol.6, p.68
31 Santa Cruz, *Reflexions militaires et politiques*, vol.6, p.78.

Years War, and the British Army of the American War of Independence. Both the Swedish and British forces were highly tactically successful in the early years of their respective conflicts. The Prussians, as a result of the heavy losses sustained in the early years of the Seven Years War, reversed course to a more firepower focused doctrine. As a result of logistical challenges as well as tactical reverses, both the Swedish and British armies were eventually defeated by their more numerous and firepower-focused opponents. Using bayonet intimidation to the near exclusion of firepower was wildly successful in the short term, but did not result in operational and strategic victory for either power. Most eighteenth-century armies were more flexible, trying to use a combination of firepower and the threat of bayonets to drive enemy battalions off the battlefield. The reality of the eighteenth-century battle lay somewhere in the dense smoke, between the officers' desires for rigid control and common soldiers' psychological instincts.

What of the larger idea of this book? Was the world of eighteenth-century infantry combat the birth of modern infantry warfare? This is a question that would require a much broader scope to definitively answer, tracing the impact that the emergence of firearms have had on common soldiers and their world of battle from 1500 to the present. As a parochial scholar of the eighteenth century, the current author must await the answer from a historian who can examine this broader period with skill. This story must be told not only across military Europe, but as a truly global history of the emergence of modern infantry.

What this book has demonstrated, however, is that infantry soldiers in the eighteenth century sometimes disobeyed their officers' commands and fought in a way which made sense to them. They aimed their weapons, fought as skirmishers, and fired independently and uncontrolled, often at a greater range than their officers preferred. They prioritized cover on the battlefield and ran at speed to avoid staying in the line of fire more than necessary. Despite their officers yearning for the reintroduction of the pike, they valued the bayonet as a weapon of intimidation, but preferred to avoid hand-to-hand combat where possible. In many ways, these eighteenth-century soldiers, rather than their officers, were the forward-thinking visionaries of the future of infantry warfare.

Bibliography

Manuscript Sources

British Manuscript Sources
Bedfordshire and Luton Archives and Record Service, Bedford
R 769, Sgt. John Harvey to the Duke of Beford
ABM1779–1782, Marriage Bonds and Allegations

Berkshire Record Office, Reading
R/D/134/13, George Dawson to his Mother
D/EN/F54, Journal of a Tour in Switzerland and Italy, 1744

Bodleian Library, Oxford
English Historical Manuscripts C 282, Narrative of General Conway's Tour

British Library, London
King's Manuscripts, 240, Sir David Dundas, Memorandum
King's Manuscripts, 241, Sir David Dundas, Remarks on the Prussian Troops and their Movements
European Manuscripts, B296, Letters and other Papers of Samuel Hickson

Dorset History Centre, Dorchester
D/WIB/C/93, Letter from Robert Honeyborne to his wife, Jane
PE-BF/OV/5, Blandford Parish, Overseers of the Poor, Settlement and Removal Papers

Dunvegan Castle Archives, Dunvegan
MacLeod Papers, Norman MacLeod, NRAS 2950/4/752, Letter from Col. McLeod to the Col. of the 73rd, 1787

Gloucestershire Archives, Gloucester
D4582 Bowly Family of Cirencester
D153 Jackson Family of Sneyd Park

Lambeth Palace Library, London
Beloe Papers, MS 3263

Lancashire Archives, Preston
DDX 2743/MS5237, William Smith to his Wife
QSP/1996/9, William Bromley to his Father

The National Archives of the United Kingdom, Kew (TNA)
FO 95/5/3, Camp near Tournay. Sir William Erskine to Dundas
HCA 30/272/3, Papers of several vessels, including *c* 100 personal and business letters for onward delivery, the largest portion of which are from British soldiers on board troopships arriving at Rhode Island in February 1777
HO 42/46/32, Letter from Isaac Richie, an old soldier, now a debtor in the Common Side, New Gaol, Southwark
SP 36/72/124, David Campbell, soldier in Capt. Buchanan's company in Gen. Colyear's regiment, at Hertogenbosch, in North Brabant
SP 36/84/2/8, to the Duke of Montague from the soldiers of his regiment, petitioning for release home 'as soon as other regiments'
SP 54/26/122, Letters from Jacobite soldiers to family and friends. Giving intelligence of their strength, movements and morale
WO 7/25, Board of General Officers, Clothing
WO 28/8/137, Copy of a letter from James Milligan, soldier in the 84th Regiment, written to a friend on the Mohawk River
WO 40/3, Selected Petitions: Innkeepers etc.
WO 97, Royal Hospital Chelsea, Soldiers Service Documents

National Records of Scotland, Edinburgh
GD248/509/3/74, Papers of the Ogilvy family, Earls of Seafield, Letter from Duncan Grant to his father

Norwich University Kreitzberg Library Special Collections, Northfield
356.10943 P972n, New Regulations for the Prussian Infantry, Signed Christopher Darby, January 9th, 1784

Royal Archives, Windsor
Cumberland Papers, Main Series.

Templer Study Centre, National Army Museum, London (NAM)
1976-07-40, Letter of James Castle, 1743
1986-11-1, Letters of Sergeant Calder, 1778-1785
1986-12-38-116, Papers of Alexander Dury
2008-06-4, Letter of Private W. Hopkins
2010-11-16, Roger Lamb Scrapbook
2012-08-2, Papers of Adam Williamson

West Sussex Record Office, Chichester
RSR/MSS/1/16, *Derby Mercury* print of a letter from a soldier in the 35th Regiment

German Manuscript Sources

Brandenburgisches Landeshauptarchiv, Potsdam-Golm (BLHA)
Rep. 2A III, Nr. D9607
Rep. 2 Kurmarkischer Kammer, Nrs: D4041, F5342, S2152, S6432 S6482/1, S6483, S6964, S7206
Rep. 3 Neumarkischer Kammer, Nrs.: 6688, 7288, 9203
Rep. 3B III, Nr. D913
Rep. 6B Soldin, Nr. 184;
Rep. 17B, Nr. 3529
Rep. 19 Lindow/M 60;
Rep. 19 Potsdam, Nrs. 1162, 1277, 1568, 1571, 1576, 2472
Rep. 23A, Nr. A 116, Nr. B 1794, Nr. C 1921;
Rep. 23B, Nr. 2936.
Rep. 37 Neuhardenburg, Nr. 1571, 'Militaria Anglica'.

Geheimes Staatsarchiv Preußischer Kulturbesitz, Berlin-Dahlem (GStAPK)
Hauptabteilung IV Rep. 15A, Nr. 610, Rélation von der Bataille von Lowositz am 1ten Oktober 1756
Hauptabteilung X, Rep. 37 Stavenow Nr. 496, Letter of Hans Wolcke

Hessisches Hauptstaatsarchiv, Wiesbaden (HHStAW)
Fonds 133, No. 11670, Letter of Philipp Reihard Weingarten

Hessissches Staatsarchiv, Darmstadt (HStAD)
Bestand D 4, Nr. 530/5, Prussian Army Files
Bestand G 28, Nr. F 2017, Letters of Philipp Wolfenstaedter
Bestand O 59, Nr. 10, Köhler Family Correspondence

United States Manuscript Sources
National Archives of the United States, Washington, D.C.
M804, Revolutionary War Pension and Bounty-Land-Warrant Application Files

Library of Congress, Washington D.C.
Washington Papers

Society of the Cincinnati Library, Washington, D.C.
MSS L1992.1.477, Gibraltar Orderly Book
MSS L2017F30, Duke of Cumberland's Orderly Book

Published Primary Sources

Agar, William, *Military Devotion: or the soldier's duty to God* (London: P. Brindley, 1758)
Adams, John Quincy, *Letters on Silesia: Written During a Tour Through that Country* (London: J Budd, 1804)
Aldersberg, Joseph Berhandtsky von, *Der Graf von Sonnetnthal oder: Das Schicksal des Soldaten* (Munich: Churfürstlichen Genehmhaltung, 1777)
Alicia, Rosalind, et al., *Report on the Manuscripts of Mrs. Franklin-Russell-Astley: of Chequers Court, Bucks* (London: Historical Manuscripts Commission, 1900)
Anon., *The Annual Register for the Year 1758* (London: Dodsley, 1764)
Anon., *The Arminian Magazine* (London: J. Fry, 1778)
Anon., *British Glory Reviv'd* (London, J. Roberts, 1743)
Anon., *The Methodist Magazine for the Year 1799, being a continuation of the Arminian Magazine* (London: G. Whitfield, 1799)
Anon., *Offizier-Lesebuch, Historisch-Militärischen Inhalts, mit untermischten interessanten Anekdoten, von einer gesellschafts militärischer Freunde* (Berlin: C. Matzdorff's Buchhandlung, 1793)
Anon., *The Parliamentary Register or History of the Proceedings and Debates of the House of Commons* (London: J. Almon, 1780)
Anon., *Sammlung ungedruckter Nachrichten* (Dresden: Waltherischen Buchhandlung, 1782)
Archenholz, Johann Wilhelm von, *England und Italien* (Leipzig: Dykischen Buchhandlung, 1787)
Barsewisch, Ernst von, *Meine Kriegs-Erlebnisse während des Siebenjährigen Krieges, 1757–1763* (Berlin: Warnsdorff, 1863)
Berenhorst, Georg Heinrich von and Eduard von Bülow, *Aus dem Nachlasse von Georg Heinrich von Berenhorst* (Dessau: K. Aue, 1845)
Berner, Ernst, and Gustav Berthold Volz, *Aus der Zeit des Siebenjährigen Krieges: Tagebuchblätter und Briefe der Prinzessin Heinrich und des Königlichen Hauses* (Berlin: Duncker, 1908)
Büschel, Johann, *Neue Reisen eines Deutschen nach und in England im Jahre 1783* (Berlin: Friedrich Mauer, 1784)
Blaufarb, Rafe, and Claudia Liebeskind, *Napoleonic Foot Soldiers and Civilians: a Brief History with Documents* (Bedford: St. Martin's, 2011)
Bleckwenn, Hans, *Preussische Soldatenbriefe* (Osnabrück: Biblio Verlag, 1982)
Boswell, James, *Journal of his Swiss and German Travels* (Edinburgh: Edinburgh University Press, 2008)
Bräker, Ulrich, *Lebensgeschichte und Natürliche Abentheuer eines Armen Mannes von Tockenburg* (Zurich: Hans Heinrich Füssli, 1789)
Cornwallis, Charles, *Correspondence of Charles, First Marquis Cornwallis* (London: John Murray, 1859)
Cogniazzo, Jakob de, *Geständnisse eines Oesterreichischen Veterans* (Breslau: Löwe, 1791)
Cope, John, et al, *The Report of the Proceedings and Opinion of the Board of General Officers, on Their Examination into the Conduct, Behaviour, and Proceedings*

of, Sir John Cope, Peregrine Lascelles, and Thomas Fowke, from the Time of the Breaking out of the Rebellion in North-Britain in the Year 1745, till the Action at Preston-Pans Inclusive (London: W. Webb, 1749)

Cranfield, Richard, *The Christian; a Memoir of Thomas Cranfield* (London: Religious Tract Society, 1844)

Dillon, John. *A Short Account of Mr. John Dillon, Preacher of the Gospel written by Himself* (London: G. Whitfield, 1796)

Dittfurth, Franz Wilhelm von, *Zehn Schöne neue Lieder aus dem Siebenjährigen Kriege* (Berlin: Trowisch und Sohn, 1851)

Dittfurth, Franz Wilhelm von, *Die historischen Volkslieder des Siebenjährigen Krieges, nebst geschichtlichen und sonstigen Erläuterungen* (Berlin: Franz Lipperheide, 1871)

Dominicus, Johann Jakob, Dietrich Kerler (ed.), *Aus dem Siebenjährigen Krieg. Tagebuch des preußischen Musketiers Dominicus* (Munich: C.B. Beck'sche Verlag, 1891)

Doddridge, Phillip, *Some Remarkable Passages in the Life of the Honourable Col. James Gardiner: Who Was Slain at the Battle of Preston-Pans, September 21, 1745* (London: Buckland et al, 1772)

Douglas, James, *Travelling Anecdotes: through Various Parts of Europe* (London: J. Debrett, 1782)

Dreyer, Jospeh Ferdinand, *Leben und Taten eines Preussischen Regiments-Tambours von ihm selbst beschrieben in seinem 93ten Lebenjahre* (Altpreussischer Komimiss 22. Osnabrück: Biblio, 1975 [1810])

Du Roi, August Wilhelm, *Journal of Du Roi the Elder* (Philadelphia: University of Pennsylvania Press, 1911)

Eckarthausen, Carl, *Erzählungen für empfindsame Herzen an Sonnabenden nach der Arbeit.* (Munich: Strohl, 1784)

Eddie, S.A. *Freedom's Price: Serfdom, Subjection and Reform in Prussia, 1648–1848* (Oxford, Oxford University Press, 2013)

Fonblanque, Edward, *Political and Military Episodes in the Latter Half of the Eighteenth Century* (London: MacMillan, 1876)

Foote, Jeffrey, *A Defence of the Planters of the West-Indies* (London: J. Debrett, 1792)

Friedrich II, *Instruktion für die Commandeurs die Infanterie-Regimenter* (Berlin: 1763)

Friedrich II, *Oeuvres de Frederic le Grand* (Berlin: 1846–1867)

Guibert, Jacques-Antoine-Hippolyte comte de, Joseph Johnson (trans.), *Observations on the military establishment and discipline of the King of Prussia* (London: Fielding and Walker, 1780)

Guibert, Jacques-Antoine-Hippolyte comte de, *Journal D'un Voyage En Allemagne, Fait En 1773* (Paris: Publisher Unknown, 1803)

Haller, Franz Ludwig von, *Militärischer Charakter und Merkwürdige Kriegsthaten Friedrich des Einzigen: Königs von Preussen: Nebst einem Anhang über einige seiner Berühmtesten Feldherren und verschiedene Preussische Regimenter* (Berlin: Bei Oemigke Dem Jüngern, 1796)

Helfferich, Tryntje, *The Thirty Years War: a Documentary History* (Indianapolis: Hackett, 2009)

Herrmann, Otto, 'Prinz Ferdinand Von Preussen über Den Feldzug Vom Jahre 1757,' *Forschungen zur brandenburgischen und preußischen*, vol. 31, (1919), pp 85–105
Hildebrandt, C., *Anekdoten und Charakterzüge aus dem Leben Friedrichs des Grossen* (Halberstadt: Brüggemannn, 1829)
Hope, John, *Hope's Curious and Comic Miscellaneous Works* (London: unknown, 1780)
Hoyer, Johann Gottfried von, *Geschichte der Kriegskunst* (Göttingen: Rosenbusch, 1797)
Jackson, Robert, *A Systematic View of the Formation, Discipline, and Economy of Armies* (London: Stockdale, 1804)
Jackson, Thomas, *The Lives of Early Methodist Preachers: Chiefly Written by Themselves* (London: Wesleyan Conference Office, 1875)
Jany, Curt, 'Briefe Preussischer Soldaten aus den Feldzügen 1756 und 1757 und über die Schlachten bei Lobositz und Prag,' in Curt Jany, *Urkundliche Beiträge Und Forschungen Zur Geschichte Des Preussischen Heeres* (Berlin: E.S. Mittler, 1901)
Kaltenborn, Rudolph Wilhelm von, *Briefe eines alten Preussischen Officiers verschiedene Characterzüge Friedrichs Des Einzigen Betreffend* (Hohenzollern: Publisher Unknown, 1790)
Kessler, Christian David, et al, *Neues Journal für Prediger* (Halle: Karl August Kummel, 1807)
Kohl, Rolf Dieter. 'Ein Brief des Wiblingwerder Bauernshones Johann Hermann Dresel aus dem Siebenjährigen Krieg, *Die Märker*, Volume 28, No. 3 (1979) pp.82–84
König, Anton Balthasar, *Biographisches Lexikon aller Helden und Militairpersonen* (Berlin: Wever, 1790)
Küster, Carl Daniel, *Des Preussischen Staatsfeldpredigers Küster, Bruchstück Seines Campagnelebens im Siebenjährigen Kriege* (Berlin: Karl Massdorfs, 1791)
Lange, Eduard, *Die Soldaten Friedrich's Des Großen* (Leipzig: Avenarius & Mendelssohn, 1852)
Lafayette, Marquis de, *Memoirs, Correspondence and Manuscripts of General Lafayette* (London: Saunders and Otley, 1837)
Liebe, Georg, *Preußische Soldatenbriefe aus dem Gebiet der Provinz Sachsen im 18. Jahrhundert* (Halle/Salle: Gebauer Schwetschke, 1912)
Lojewsky, J. G., *Selbstbiographie des Husaren-Obersten von... ly; Oder, Meine Militairische Laufbahn im Dienste Friedrichs des Einzigen* (Leipzig: Kollmann, 1843)
Lossow, Ludwig von, *Denkwürdigkeiten zur Charakteristik der Preussischen Armee unter dem Grossen König Friedrich Dem Zweiten* (Glogau: Carl Heymann, 1826)
Mauvillon, Éléazar de, *The History of Prussia, Particularly During the Reign of the Late King Frederick William* (London: R. Manby, 1756)
Mercoyrol, Jacques de, *Campagnes De Jacques De Mercoyrol De Beaulieu, Capitaine Au régiment De Picardie (1743–1763)* (Paris: Renouard, H. Laurens, 1915)
Mirabeau, H. G. and J. Mauvillion, *Systeme Militaire de la Prusse et Principes de la Tactique Actuelle des Troupes les plus Perfectionnees. Extrait de la Monarchie Prussienne* (London: Publisher Unknown, 1788)
Miranda, Fransico de, *Archivo del General Miranda* (Caracas: Parra Leon Hermonas, 1929–1950)

Moore, John, *A View of Society and Manners in France, Switzerland, and Germany* (London: Strahan and Cadell, 1770)

Morris, Margaret, *Private Journal Kept During a Portion of the Revolutionary War, for the Amusement of a Sister* (Philadelphia: Publisher Unknown, 1836)

Moritz, Carl P, *Reisen eines Deutschen in England im Jahr 1782 in Briefen an Herrn Oberkonsistorialath Gedike* (Berlin: Friedrich Maurer, 1785)

Parkinson, James, *The Soldier's Tale, Extracted from the Village Association* (London: Eaton, 1793)

Pollnitz, Karl Ludwig, *Memoirs of Charles Louis, Baron de Pollnitz* (London: Daniel Browne, 1737)

Preuss, Johann D.E., *Urkundenbuch zu der Lebensgeschichte Friedrichs des Großen* (Berlin: Nauck, 1832)

Pouchot, Pierre, Michael Cardy (trans.), Brian Leigh Dunnigan (ed.), *Memoirs of the Late War in North America between France and England* (Youngstown: Old Fort Niagara Association, 2004)

Reisbeck, Johann K, *Briefe eines reisenden Franzosen über Deutschland an seinen Bruder zu Paris* (Location Unknown: K. R. Press 1784)

Retzow, Friedrich A, von, *Charakteristik der Wichtigsten Ereignisse des Siebenjährigen Krieges, in Rücksicht auf Ursachen und Wirkungen* (Berlin: Himburg, 1802)

Schimmel, Johann Christian, 'Kurze Lebensbeschreibung des preussischen Veterans Johann Christian Schimmel,' *Zeitschrift für Kunst, Wissenschaft, und Geschichte des Krieges*, vol.10, no.4–6 (1827), pp.188–201

Sidney, Edwin, *The Life, Ministry, and Selections from the Remains of the Rev. Samuel Walker* (London: Seeley and Burnside 1838)

Seipp, Christoph, *Reisen von Preßburg durch Mähren, beyde Schlesien u. Ungarn nach Siebenbürgen* (Frankfurt und Leizpig: Publisher Unknown, 1793)

Storring, Adam, 'Pastor Täge's Account of the Siege of the Cüstrin and the Battle of Zorndorf,' in Alexander S. Burns (ed.), *The Changing Face of Old Regime Warfare* (Warwick: Helion, 2022)

Telford, John, *Wesley's Veterans: Lives of Early Methodist Preachers* (Salem, OH: Schmul Publishers, 1912)

Trenchard, John and Thomas Gordon, *A Collection of Tracts by the Late John Trenchard* (London: Cogan, 1751)

Vattel, Emmerich de, Joseph Chitty (trans.), *The Law of Nations, or Principles of the Law of Nature* (Philadelphia: Johnson, 1844)

Volz, Gustav Berthold, *Die Politischen Testamente Friedrich's Des Großen* (Berlin: Hobbing, 1920)

Walter, Jakob, and Marc Raeff, *The Diary of a Napoleonic Foot Soldier* (Moreton: Windrush, 1999)

Washington, George, Philander D. Chase and Frank E. Grizzard, Jr (ed.) *The Papers of George Washington. Revolutionary War Series, Volume 6, 13 August 1776–20 October 1776.* (Charlottesville: University Press of Virginia, 1994)

Wesley, John, Thomas Jackson (ed.), *The Journal of the Rev. John Wesley, A.M.* (London: Wesleyan Conference Office, 1869)

Witzleben, Gerhard August von, *Aus Alten Parolebüchern der Berliner Garnison zur Zeit Friedrichs des Grossen* (Berlin: Mittlers, 1851)

Zander, Christian F., *Fundstücke – Dokumente und Briefe einer Preußischen Bauernfamilie: (1747-1953)* (Hamburg: Kovacĭ, 2015)

Zimmermann, Johann Georg, *Ueber Friedrich den Grossen* (Wien: Ofen, 1788)

Secondary Sources

Published Sources

Abraham William J. and James E. Kirby (eds), *The Oxford Handbook of Methodist Studies* (Oxford: Oxford University Press, 2009)

Ancell, Samuel. *A Circumstantial Journal of the Long and Tedious Blockade and Siege of Gibraltar from the 12th of September, 1779, to the 23d. of February, 1783, Etc* (Cork: A. Edwards, 1793)

Anderson, Fred, *A People's Army: Massachusetts Soldiers and Society in the Seven Years' War* (Chapel Hill: University of North Carolina Press, 1984)

Anderson, Fred, *Crucible of War: The Seven Years War and the Fate of Empire in British North America, 1754–1766* (New York: Alfred A Knopf, 2000)

Anderson, M.S., *War and Society in Europe of the Old Regime* (London: Leicester University Press, 1988)

Anderson, M.S., *The War of Austrian Succession 1740–1748* (London: Routledge, 1995)

Babel, Rainer et al (eds), *Grand Tour: Adeliges Reisen und europäische Kultur vom 14. Bis zum 18. Jahrhundert: Akten der Internationalen Kolloquien in der Villa Vigoni 1999 und im Deutschen Historischen Institut Paris 2000* (Ostfildern: Thorbecke, 2005)

Babits, Lawrence, *A Devil of Whipping: The Battle of Cowpens* (Chapel Hill: University of North Carolina Press, 2001)

Babits, Lawrence, et al, *Long, Obstinate and Bloody: The Battle of Guilford Courthouse* (Chapel Hill: University of North Carolina Press, 2009)

Badone, Giovanni Cerino, *You Have to Die in Piedmont! The Battle of Assietta, 19 July 1747 and the War of Austrian Succession in the Alps* (Warwick: Helion & Co, 2023)

Berkovich, Ilya, 'Fear Honour, and Emotional Control on the Eighteenth Century Battlefield,' in Erika Kuijpers and Cornelius van der Haven (eds), *Battlefield Emotions: 1500-1800, Practices, Experience, Imagination* (London: Palgrave Macmillan, 2016)

Berkovich, Ilya, *Motivation in War: The Experience of Common Soldiers in Old-Regime Europe* (Cambridge: Cambridge University Press: 2017)

Bernhardi, Theodor von, *Friedrich der Grosse als Feldherr* (Berlin: E.S. Mittler und Sohn, 1881)

Biddle, Stephen, *Military Power: Explaining Victory and Defeat in Modern Battle* (Princeton: Princeton University Press, 2004)

Birgfeld, Johannes, *Krieg und Aufklärung: Studien zum Kriegsdiskurs in der deutschsprachigen Literatur des 18. Jahrhundert* (Hannover: Wehrhahn, 2012)

Blackbourn, David and Geoff Eley, *The Peculiarities of German History: Bourgeois Society and Politics in Nineteenth-Century Germany* (Oxford: Oxford University Press, 1984)

Blackmore, David, *Destructive and Formidable: British Infantry Firepower, 1642-1765* (London: Frontline Books, 2014)
Blanning, Timothy, *The Pursuit of Glory: Five Revolutions that Made Modern Europe,1648-1815* (London: Penguin, 2008)
Blanning, Timothy, *Frederick the Great: King of Prussia* (Random House: New York, 2016)
Boer, Roland. 'EP Thompson and the Psychic Terror of Methodism,' *Thesis Eleven* 110, no. 1 (2012) pp.54–67
Bohls, Elizabeth A. et al (eds), *Travel Writing, 1700-1830: an Anthology* (Oxford: Oxford University Press, 2008)
Boland, Irene et al, *Eutaw Springs: The Final Battle of the American Revolution's Southern Campaign* (Columbia: University of South Carolina Press, 2017)
Bonomi, Patricia U., *Under the Cope of Heaven* (New York: Oxford University Press, 1986)
Bowler, Arthur, *Logistics and the Failure of the British Army in America, 1775–1783* (Princeton: Princeton University Press, 1975)
Brewer, John. and Eckhart Hellmuth, *Rethinking Leviathan: The Eighteenth-Century State in Britain and Germany* (Oxford: Oxford University Press, 1999)
Browning, Reed, *The War of the Austrian Succession* (St. Martin's: Griffin, 1993)
Büsch, Otto, *Militärsystem und Sozialleben im alten Preußen: Die Anfänge der sozialen Militarisierung der preußisch-deutschen Gesellschaft 1713–1807* (Berlin: Walter de Gruyter, 1962)
Büsch, Otto, John G. Gagliardo (trans.), *Military System and Social Life in Old Regime Prussia, 1713-1807 the Beginnings of the Social Militarization of Prusso-German Society* (Boston: Humanities Press, 1997)
Butler, Jon, *Awash in a Sea of Faith* (Cambridge: Harvard University Press, 1990)
Campbell, Alexander, *The Royal American Regiment: An Atlantic Microcosm* (Norman: University of Oklahoma Press, 2010)
Carlyle, Thomas, *History Friedrich II of Prussia, Called Frederick the Great* (London: Estes and Lauriat, 1858)
Charters, Erica et al, *Civilians and War in Europe, 1618–1815* (Liverpool: Liverpool University Press, 2014)
Childs, John, *Armies and Warfare in Europe: 1648–1789* (Manchester: Manchester University Press, 1989)
Childs, John, 'The Army and the State in Eighteenth-Century Britain and Germany,' in John Brewer and Eckhart Hellmuth (eds), *Rethinking Leviathan: The Eighteenth-Century State in Britain and Germany* (Oxford: Oxford University Press, 1999)
Churchill, Winston, *A History of the English Speaking Peoples: The Age of Revolution* (London: Cassell, 1956)
Clark, Christopher, *Time and Power: Visions of History in German Politics, from the Thirty Years War to the Third Reich* (Princeton: Princeton University Press, 2019)
Cleare. G.H, 'County Names for the regiments in 1782,' *Journal of the Society for Army Historical Research*, vol.36, no.145 (March 1958), pp.34–38
Clodfelter, Micheal, *Warfare and Armed Conflicts: a Statistical Encyclopaedia of Casualty and Other Figures, 1492–2015* (Jefferson: McFarland & Company, 2017)

Colley, Linda, *Britons: Forging the Nation, 1707–1837* (London: Pimlico, 2003)
Collins, Kenneth J., *A Wesley Bibliography* (Wilmore: First Fruits Press, 2017)
Conway, Stephen, 'The British Army, "Military Europe," and the American War of Independence', *The William and Mary Quarterly*, vol.67, no.1 (2010), pp.69–100
Conway, Stephen, 'Moral Economy, Contract, and Negotiated Authority in American, British, and German Militaries, ca. 1740-1783', *The Journal of Modern History*, vol.88, no.1 (March 2016), pp.34–59
Cormack, Andrew E, *'These Meritorious Objects of the Royal Bounty': The Chelsea Out-Pensioners in the Early Eighteenth Century* (London: Henry Ling, 2017)
Delbrück, Hans, *Die Strategie des Perikles erläutert durch die Strategie Friedrichs des Grossen: Mit einem Anhang über Thucydides und Kleon* (Berlin: Georg Reimer Verlag. 1890)
Devine, Thomas, *The Scottish Clearances: A History of the Dispossessed, 1600–1900* (London: Allen Lane, 2018)
Duffy, Christopher, *The Army of Frederick the Great* 1st Edition. (London: Hippocrene Books, 1974)
Duffy, Christopher, *Fire and Stone: The Science of Fortress Warfare, 1660-1800* (London: Peters, Fraser, and Dunlop, 1975)
Duffy, Christopher, *Russia's Military Way to the West: Origins and Nature of Russian Military Power, 1700-1800* (Routledge: London, 1981)
Duffy, Christopher, *The Military Experience in the Age of Reason* (New York: Atheneum, 1988)
Duffy, Christopher, *The Army of Frederick the Great*. 2nd Edition (Chicago: Emperor's Press, 1994)
Duffy, Christopher, *Instrument of War: The Austrian Army in the Seven Years War* (Chicago: Emperor's Press, 2000)
Duffy, Christopher, *Prussia's Glory: Rossbach and Leuthen* (Chicago: Emperor's Press, 2003)
Duffy, Christopher, *By Force of Arms: The Austrian Army in the Seven Years War* (Chicago: Emperor's Press, 2008)
Duffy, Christopher, *Fight for a Throne: The Jacobite '45 Reconsidered* (Solihull: Helion, 2015)
Duffy, Christopher, *The Army of Frederick the Great*. Revised 2nd Edition (Warwick: Helion, 2022)
Dwyer, Phillip, 'It Still Makes Me Shudder': Memories of Massacres and Atrocities during the Revolutionary and Napoleonic Wars', *War in History*, vol.16 Issue 4, (2009) pp.381–405
Dwyer, Phillip, 'Violence and the revolutionary and Napoleonic wars: massacre, conquest and the imperial enterprise,' *Journal of Genocide Research*, vol.15, vo.2, (2013), pp.117–131
Eckhardt, William, 'Civilian Deaths in Wartime,' *Bulletin of Peace Proposals* 20, no.1 (1989), pp.89–98
Eglin, John, *Venice Transfigured: the Myth of Venice in British Culture, 1660-1797* (New York: Palgrave, 2001)
Engelen, Beate, *Soldatenfrauen in Preußen: Eine Strukturanalyse der Garnisonsgesellschaft im späten 17. und im 18. Jahrhundert* (Münster: Lit Verlag, 2005)

Engelen, Beate, "'Fremde in Der Stadt: Die Garnisongesellschaft Prenzlaus Im 18. Jahrhundert,'" in Jürgen Theil, Olaf Gründel, and Klaus Neitmann (eds), *Die Herkunft Der Brandenburger: Sozial- Und Mentalitätsgeschichtliche Beiträge Zur Bevölkerung Brandenburgs Vom Hohen Mittelalter Bis Zum 20. Jahrhundert* (Potsdam, 2001), pp.116–120

Englund, Peter, *The Battle That Shook Europe: Poltava and the Birth of the Russian Empire* (London: I.B. Tauris, 2013)

Ferling, John, *Almost a Miracle: The American Victory in the War of Independence* (New York: Oxford University Press, 2007)

French, David, *Military Identities: The Regimental System, the British Army, and the British People, 1870-2000* (New York: Oxford University Press, 2008)

Frey, Sylvia, *The British Soldier in North America: A Social History of Military Life in the Revolutionary Period* (Austin: University of Texas Press, 1981)

Fiedler, Siegfried, *Kriegswesen und Kriegführung im Zeitalter der Kabinettskriege* (Koblenz: Bernard & Graef, 1986)

Forrest, Alan, *Napoleon's Men: The Soldiers of the Revolution and Empire* (New York: Continuum International Publishing, 2006)

Füssel, Marian, 'Féroces et barbares? Cossacks, Kalmyks, and Russian Irregular Warfare during the Seven Years War,' in Mark Danely and Patrick Speelman (eds), *The Seven Years War: Global Views* (Boston: Brill, 2012)

Füssel, Marian, 'Emotions in the Making: The Transformation of Battlefield Experiences during the Seven Years War' in Erika Kuijpers and Cornelius van der Haven (eds), *Battlefield Emotions 1500-1800: Practices, Experience, and Imagination* (London: Palgrave Macmillan, 2016)

Füssel, Marian, *Der Preis des Ruhms: Eine Weltgeschichte des Siebenjährigen Krieges, 1756–1763* (Munich: C.H. Beck, 2019)

Gawthrop, Richard L., *Pietism and the Making of Eighteenth-Century Prussia* (Cambridge: Cambridge University Press, 1993)

Gooch, George P., *History and Historians in the Nineteenth Century* (London: Longman, 1913)

Griep, Wolfgang et al (eds), *Reisen im 18. Jahrhundert: Neue Untersuchungen* (Heidelberg: C. Winter Universitätsverlag, 1986)

Grimsley, Mark, and Clifford Rogers, *Civilians in the Path of War* (Lincoln: University of Nebraska Press, 2002)

Gritsch, Eric, 'Luther and the State: Post-Reformation Ramifications,' in James Tracey (ed.), *Luther and the Modern State in Germany* (Kirksville: Missouri, 1986)

Gruber, Ira D., *Books and the British Army: in the Age of the American Revolution* (Chapel Hill: University of North Carolina Press, 2010)Covers

Hagen, William W., *Ordinary Prussians: Brandenburg Junkers and Villagers 1500-1840* (Cambridge: Cambridge University Press, 2002)

Hagist, Don, *British Soldiers, American War: Voices of the American Revolution* (Yardley: Westholme, 2014)

Hagist, Don, *Noble Volunteers: The British Soldiers who Fought the American Revolution.* (Yardley: Westholme, 2020)

Hamilton, Phillip and Glenn Moots, *Justifying Revolution: Law, Virtue, and Violence in the American War of Independence* (Norman: University of Oklahoma Press, 2018)

Harari, Yuval N., *Ultimate Experience: Battlefield Revelations and the Making of Modern War Culture, 1450-2000* (London: Palgrave Macmillan, 2014)

Hayter, Anthony J., *The Army and the Crowd in Mid-Georgian England* (London: The MacMillan Press, 1978)

Heathorn, Stephen, 'E.P. Thompson, Methodism, and the "Culturalist" Approach to the Historical Study of Religion,' *Method & Theory in the Study of Religion*, vol.10, no.2 (1998), pp.210–226

Hempton, David, *Methodism and Politics in British Society, 1750-1850* (London: Routledge, 1984)

Heyden, Hellmuth, 'Die Kirchenpolitik in Pommern von der Teilung des Landes 1648 bis zur Mitte des 19. Jahrhunderts.' *Baltische Studien*, no.57 (1971), pp.51–65

Hinrichs, Carl, *Preussentum und Pietismus: Der Pietismus in Brandenburg-Preussen als Religiössozial Reformbewegung* (Göttingen: Vandenhoeck & Ruprecht, 1971)

Hintze, Otto, 'Die Epochen Des Evangelischen Kirchenregiments in Preussen,' in *Regierung und Verwaltung: Gesammelte abhandlungen zur Staats-, Rechts- und Sozialgeschichte Preussens* (Göttingen: Vandenhoeck und Ruprecht, 1967), pp.56–96

Hintze, Otto, *The Historical Essays of Otto Hintze* (New York: Oxford University Press, 1975)

Hohrath, Daniel, *The Uniforms of the Prussian Army under Frederick the Great* (Berlin: Militaria Verlag, 2011)

Hoock, Holger, *Scars of Independence: America's Violent Birth* (New York: Crown, 2017)

Houlding, John, *Fit for Service: The Training of the British Army, 1715–1795* (Oxford: Oxford University Press, 1981)

Howard, Michael. *War in European History* (New York: Oxford University Press, 2009)

Hulme, Peter, et al (eds), *The Cambridge Companion to Travel Writing* (Cambridge: Cambridge University Press, 2013)

Iida, Takashi. 'Coping with Poverty in Rural Brandenburg: The Role of Lords and State in the Late Eighteenth Century', in Tanimoto Masayuki and Wong R. Bin (eds), *Public Goods Provision in the Early Modern Economy: Comparative Perspectives from Japan, China, and Europe* (Oakland, California: University of California Press, 2019)

Kloosterhuis, Jürgen, *Bauern, Bürger und Soldaten. Quellen zur Sozialisation des Militärsystems im preußischen Westfalen 1713–1803* (Münster: De Gruyter, 1992)

Kloosterhuis, Jürgen, 'Donner, Blitz und Bräker – der Soldatendienst des armen Mannes im Tockenburg aus der Sicht des preußischen Militärsystems', in *Schreibsucht – autobiografische Schriften des Pietisten Ulrich Bräker (1725–1798)* (Göttingen: Vandenhoeck & Ruprecht, 2004)

Kloosterhuis, Jürgen, *Militär und Gesellschaft in Preußen-Quellen zur Militärsozialisation, 1713–1806* (Frankfurt: Peter Lang, 2014)

Kocka, Jurgen, 'German History before Hitler: The Debate about the German Sonderweg', *Journal of Contemporary History*, vol.23, no.1 (1988), pp.3–16

Koser, Reinhold, *Geschichte Friedrichs des Großen* (Stuttgart & Berlin: J.G. Cotta, 1921)

Landa, Manuel de, *War in the Age of Intelligent Machines* (New York, NY: The MIT Press, 1991)

Charles T. Lanham, *Infantry in Battle*, 2nd Edition (Richmond: Garrett & Massie, 1939)

Langford, Paul, *A Polite and Commercial People: England, 1727-1783* (Oxford, 1989: Oxford University Press, 1989)

Layne, Darren, 'Layne on Massie and Oates,' *The Scottish Historical Review*, vol.99, no.2, (October 2020), pp.316–318

Laslett, Peter, *The World We Have Lost: Further Explored* (London: Routledge, 2004)

Leggiere, Michael V., *Blücher: Scourge of Napoleon* (Norman, University of Oklahoma Press: 2014)

Leggiere, Michael V., 'Napoleon and the Strategy of the Single Point,' in Hal Brands (ed.), *The New Makers of Modern Strategy* (Princeton: Princeton University Press, 2023)

Leibetseder, Mathis, *Kavalierstour – Bildungsreise – Grand Tour: Reisen, Bildung Und Wissenserwerb in Der Frühen Neuzeit* (Mainz: Leibniz-Institut für Europäische Geschichte, 2013)

Levy, Jack S., *War in the Modern Great Power System: 1495–1975* (Lexington, KY: University Press of Kentucky, 1983)

Linn, Edward, 'The Battle of Culloden 16 April 1746 as described in a Letter from a soldier in of the Royal Army to His Wife,' *Journal of the Society for Army Historical Research*, vol.1 (1921–2), pp.21–24

Lynn, John, *The Bayonets of the Republic Motivation and Tactics in the Army of Revolutionary France, 1791–94* (Boulder: Westview Press, 1996)

Macaulay, Thomas B., *The History of England from the Accession of James the Second* (London: 1848)

Mackesy, Piers. *The War for America, 1775–1783* (Lincoln: University of Nebraska Press, 1964)

Marschke, Benjamin, *Absolutely Pietist: Patronage, Factionalism, and State-Building in the Early Eighteenth-Century Prussian Army Chaplaincy* (Leipzig: Max Niemeyer Verlag, 2005)

Marshall, P.J., *The Making and Unmaking of Empires: Britain, India, and America, c.1750-1783* (Oxford: Oxford University Press, 2005)

McCormack, Mathew, 'Citizenship, Nationhood, and Masculinity in the Affair of the Hanoverian Soldier, 1756', *The Historical Journal*, vol.49, no.4 (2006), pp.971–993

McKay, Derek, *The Great Elector* (Longman: New York, 2001)

Metcalf, Thomas, *Ideologies of the Raj* (Cambridge, Cambridge University Press 1995)

Mikaberidze, Alexander, *The Napoleonic Wars: a Global History* (Oxford: Oxford University Press, 2020)

Mingay, G.E., *English Landed Society in the Eighteenth Century* (London: Routledge: 1963)

Mintzker, Yair, *The Defortification of the German City, 1689–1866* (Cambridge: Cambridge University Press, 2012)

Mallinckrodt, Rebecca von, 'There Are No Slaves in Prussia?' in Brahm Felix, Rosenhaft Eve, Bredeck Elizabeth (eds), *Slavery Hinterland: Transatlantic Slavery and Continental Europe, 1680-1850* (Woodbridge & Rochester: Boydell and Brewer, 2016)

Miakinkov, Eugene, *War and Enlightenment in Russia: Military Culture in the Age of Catherine II* (Toronto: University of Toronto Press, 2020)

Möbius, Katrin and Sascha Möbius, *Prussian Army Soldiers and the Seven Years War: The Psychology of Honour* (London: Bloomsbury Academic, 2020)

Möbius, Sascha, *Mehr Angst vor dem Offizier als vor dem Fiend? Eine mentalitätsgeschichtliche Studie zur preußischen Taktik im Siebenjährigen Krieg* (Saarbrücken: VDM Verlag, 2007)

Möbius, Sascha, 'Kriegsgreuel in den Schlachten des Siebenjährigen Krieges in Europa,' in: Sönke Neitzel and Daniel Hohrath (eds), *Kriegsgreuel: Die Entgrenzung der Gewalt in Kriegischen Konflikten vom Mittelalter,* (Schöningh: Brill, 2008)

Muth, Jörg, *Flucht aus dem Militärischen Alltag: Ursachen und Individuelle Ausprägung der Desertion in der Armee Friedrichs des Grossen: Mit Besonderer Berücksichtigung der Infanterie-Regimenter der Potsdamer Garnison* (Freiburg Im Breisgau: Rombach Verl., 2003)

O'Shaunessy, Andrew J., *The Men who Lost America: British Leadership, the American Revolution, and the Fate of Empire* (New York: Yale University Press, 2014)

Oates, Jonathan. *Sweet William or the Butcher? The Duke of Cumberland and the '45* (London: Pen and Sword Military, 2008)

Ohlmeyer, Jane, 'A Laboratory for Empire? Early Modern Ireland and English Imperialism.' in Kevin Kenny (ed.), *Ireland and the British Empire* (New York: Oxford University Press, 2006)

Planert, Ute, *Der Mythos vom Befreiungskrieg: Frankreichs Kriege und der deutsche Süden: Alltag-Wahrnehmung-Deutung, 1792–1841* (Schöningh: Ferdinand, 2007)

Plank, Geoffrey, *An Unsettled Conquest: The British Campaign Against the Peoples of Acadia* (Philadelphia, University of Pennsylvania Press, 2003)

Paoletti, Ciro, 'War, 1688–1812,' essay, in Peter H. Wilson (ed.), *A Companion to Eighteenth -Century Europe* (Oxford: Wiley Blackwell, 2014)

Phillips, Jason, 'Battling Stereotypes: A Taxonomy of Common Soldiers in Civil War History,' *History Compass*, no. 6 (2008), pp.1417–1425

Phillips, Jason, *Diehard Rebels: The Confederate Culture of Invincibility* (Athens: University of Georgia Press, 2010)

Pichichero, Christy, *The Military Enlightenment: War and Culture in the French Empire from Louis XIV to Napoleon* (Cornell: Cornell University Press, 2017)

Pittock, Murray, *The Myth of the Jacobite Clans: The Jacobite Army in 1745* (Edinburgh: Edinburgh University Press, 2009)

Porter, Roy, *English Society in the Eighteenth Century* (London: Pelican Books, 1982)

Reid, Stuart, *1745: A Military History of the Last Jacobite Uprising* (Staplehurst: Spellmount, 1996)

Ribas, Alberto Raúl Esteban, *The Battle of Nördlingen 1634: The Bloody Fight between Tercios and Brigades* (Warwick: Helion & Co, 2021)
Ritter, Gerhard, *Europa Und Die Deutsche Frage* (München: F. Bruckmann Verlag, 1948)
Ritter, Gerhard, *Staatskunst Und Kriegshandwerk: Das Problem Des 'Militarismus' in Deutschland* (München: Oldenbourg, 1970)
Robitschek, Norbert, *Hochkirch: Eine Studie* (Wien: C. Teufen's Nachfolger, 1905)
Rodger, Nicholas A.M., *Command of the Oceans: A Naval History of Britain, 1649-1815* (London, W.W. Norton, 2005)
Salisch, Marcus von, *Treue Deserteure: Das Kursächsische Militär und der Siebenjährige Krieg* (München: R. Oldenbourg, 2009)
Schui, Florian, *Rebellious Prussians: Urban Political Culture under Frederick the Great and his Successors* (Oxford: Oxford University Press, 2013)
Schumann, Matt, 'The end of the Seven Years' War in Europe,' in Mark Danely and Patrick Speelman (eds.), *The Seven Years War: Global Views* (Boston: Brill, 2012)
Scott, Hamish, 'The Seven Years War and Europe's Ancien Regime,' *War in History*, vol.18, no.4 (2011), pp.419-455
Sikoura, Michael, *Discinplin und Desertion: Strukturprobleme militärischer Organisation im 18. Jahrhundert* (Berlin: Duncker und Humboldt, 1996)
Spring, Matthew H., *With Zeal and with Bayonets Only: the British Army on Campaign in North America, 1775-1783* (Norman: University of Oklahoma Press, 2010)
Snape, Michael F., *The Redcoat and Religion: the Forgotten History of the British Soldier from the Age of Marlborough to the Eve of the First World War* (London: Routledge, 2008)
Smith, Hannah, 'Politics, Patriotism, and Gender: The Standing Army Debate on the English Stage', *Journal of British Studies*, vol.50, (2011), pp.48-75
Smith, Hannah, 'The Army, Provincial Urban Communities, and Loyalist Cultures in England, *Journal of Early Modern History*, vol.15, (2011), pp.139-158
Sumner, Percy, 'General Hawley's Chaos,' *Journal of the Society for Army Historical Research*, vol.26, no.107, (Autumn, 1948), pp.91-94
Szabo, Franz A.J., *The Seven Years War in Europe, 1756-1763* (London: Pearson Longman, 2008)
Tatum, William P. III, '"The Soldiers Murmured much on Account of this Usage": Military Justice and Negotiated Authority in the Eighteenth-Century British Army' in Kevin Linch and Matthew McCormack (eds), *Britian's Soldiers: Rethinking War and Society, 1715-1815* (Liverpool: Liverpool University Press, 2014)
Thompson, E. P., *The Making of the English Working Class* (New York: Pantheon Books, 1964)
Thompson, E. P., 'The Moral Economy of the English Crowd in the Eighteenth Century', *Past & Present*, no.50 (1971), pp. 76-136
Wehler, Hans-Ulrich, *Preussen ist wieder chic...: Politik und Polemik in Zwanzig Essays* (Frankfurt am Main: Suhrkamp, 1983)
Wilson, Peter H., *German Armies: War and German Politics, 1648-1806* (London: University College London Press, 1998)

Wilson, Peter H., 'Prusso-German Social Militarization Reconsidered,' in Jürgen Luh, Vincent Czech, and Bert Becker (eds), *Preussen, Deutschland, Und Europe 1701–2001* (Groningen: Institute for North and East European Studies, 2003)

Wilson, Peter H., *The Thirty Years War: Europe's Tragedy* (Cambridge: Belknap Press of Harvard University Press, 2011).

Winter, Martin, *Untertanengeist durch Militärpflict: Das Preussische Kantonsystem in brandenburgischen Städten im 18. Jahrhundert* (Bielefeld: Regionalegeschichte, 2005).

Unpublished Dissertations

Backlund, Janne. 'Rusthållarna i Fellingsbro 1684–1748 : indelningsverket och den sociala differentieringen av det svenska agrarsamhället.' Ph.D diss. University of Uppsala, 1993

King, Stephen. 'The 1782 British Army Reforms: Adoption of County Titles.' MA thesis. University of Plymouth, 2011

Osman, Julia. 'The Citizen Army of Old Regime France.' Ph.D diss. University of North Carolina-Chapel Hill, 2010

Pimlott, John L. 'The Administration of the British Army, 1783–1793' Ph.D diss. Leicester University, 1975

Röder, Tobias, 'Professional Identity of Army Officers in Britain and the Habsburg Monarchy, 1740-1790.' PhD diss. University of Cambridge, 2019

Steppler, Glenn A. 'The Common Soldier in the Reign of George III, 1760-1793.' PhD diss. University of Oxford, 1984

Storring, Adam. 'Frederick the Great and the Meanings of War, 1730-1755' Ph.D diss. University of Oxford, 2018

Index

Aimed fire 36, 117–123, 135–136
American War of Independence 39, 47–48, 52–53, 57, 84–85, 90, 100, 102, 116, 118, 121, 126–127, 131–132, 135, 140–142, 147, 150, 163–167, 170, 172, 177–178, 203
Ammunition usage 46, 103, 148–151, 187–188, 195
Arms and accoutrements 41
Automata, soldiers as 76, 80–82, 84, 88, 100, 150, 166

Battle, soldiers' experience of 44–46, 197–201
Bayonet fighting *see close combat*
Brandywine 140, 163, 165, 173, 179–180
Broglie, François-Marie, duc de 18, 29, 126–127
Burgoyne, Lieutenant General John 43, 84–85, 94, 97, 132, 147

Cadenced marching 35, 49, 53–54, 88–90, 97
Camden 93, 122, 141, 179, 192
Cavalry, infantry fighting against 132–136, 168–169
Cavan, Major General Richard Lambert, the Earl of 55–56
Charging 55, 57, 94–98, 102–103, 133, 147
Charles Emmanuel III, Duca di Savoia and King of Sardinia 18–20, 25, 28, 32, 34–35
Charles VI, Emperor 19, 28
Charles XII, King of Sweden 25, 34, 95–96
Clifton 157–158
Close combat 24–25, 33–35, 37, 54, 56, 59, 95–97, 103, 141–142, 150, 157–158, 166–173, 178, 184–185, 193, 200, 202–203
Coigny, François de Franquetot, duc de 18, 36
Columns, assault 173–176, 180
Common soldiers, Prussian 64–69, 76–79, 107–115; Jacobite 69–72; British 72–76
Cover, the use of 22–23, 27, 45, 123, 157–163, 185–186
Cowpens 56, 141, 165, 167, 179, 190, 195
Culloden 70, 167, 171

Denain 18, 20
Deployiren, deployment 50–51, 54, 88

Dettingen 75, 92, 139, 142–144, 159
Discipline 47, 64, 76, 80–85, 90, 94, 97, 127, 142–143, 145
Drill 42, 48–49, 53, 85–85–87

Eugene, Prince of Savoy 18–19, 28
Eutaw Springs 56, 97, 172, 176–179, 189–191, 195–196

Fire control 33, 45–46, 58, 90–94, 96–98, 117-123, 155, 187; firings 26–27, 57–59, 97, 131, 142, 147
Firefights 23, 25, 27, 32–34, 36, 46, 59, 95, 98, 119, 132, 137–138, 140–142, 147, 150–151, 156, 158, 166, 177, 184, 187–189; ranges of 33, 36, 41, 45–46, 55–58, 91–94, 122, 138–142, 202; rate of fire 25, 26, 152–155
Firepower 53, 55–59, 90, 92–94, 97–98, 117, 122, 142, 149, 151–152, 161, 163, 167, 170, 178, 185, 189, 194, 202–203
Flanquers *see skirmishing*
Fontenoy 149, 160
Formations 49–50
Frederick II, King of Prussia 36–37, 51, 63–65, 68–69, 79, 81–83, 91, 95, 99, 114–115, 123, 125, 146, 148, 152, 155, 164–165, 169
Freeman's Farm 94, 97, 150, 166, 183
French and Indian War 92, 140

Germantown 40, 51, 56, 122, 141, 150, 176–189, 195–196
Great Northern War 19, 25, 55, 90, 141, 167, 202
Greene, Major General Nathanael 98, 179–180, 183, 189–196
Gross-Jägersdorf 51, 144
Guastalla 17, 28–38, 55, 96, 107, 132, 136, 139, 144, 159, 161, 176–178, 189, 196
Guibert, Jacques-Antoine-Hippolyte, comte de 36, 48, 57, 78, 86, 88–89, 120, 145, 167

Henry, Prince of Prussia 40, 65, 99, 105
Hochkirch 69, 139, 149, 170, 173–174, 180–181

INDEX

Hohenfriedberg 81, 161, 168

Independent fire 26–27, 36, 142–148
Itzenplitz Regiment (IR 13) 38, 65, 69, 77, 99–100, 137–138, 147

Jacobite Rebellion (1745) 69–76, 79, 114, 157

Kesselsdorf 81, 105, 140, 172
Königsegg-Rothenfels, Lothar Joseph Dominik von 19, 28–30, 32, 36

Leadership in battle 34, 46, 99–107
Leuthen 68, 139, 149
Line of battle 45, 50–52, 137, 139, 174, 177–178, 180, 192
Lobositz 51, 109, 111, 137, 147–148, 154–155
Long Island 127–128, 140, 155, 165, 179
Louvigny, marquis de 20, 27

Mantua 18–19, 28
Maxen 173–175
Mercy, Claude Florimond de 19, 22–23, 30, 112
Melee *see close combat*
Mollwitz 36, 81, 124, 138–139, 142, 148, 154
Monmouth 120, 165, 179
Moritz, Prinz of Anhalt-Dessau 103, 105, 149, 172
Movement, speed of 35–37, 46, 143, 147–150, 157, 159, 163–166, 168, 172, 185, 189, 198

Negotiated authority 24, 37, 76–77, 79, 98–102, 106–107, 115–116, 202
Nine Years War 18, 25

Paoli 170, 180, 184
Parma 17, 19–28, 30, 32–33, 35–38, 105, 132, 136, 144, 155, 159, 161, 176–178, 189, 196
Platoons 47–48
Platoon fire *see fire control, firings*
Philippsburg 18
Prague 99–100, 105, 109–110, 139, 142, 164
Princeton 104, 140, 179
Prussomania 79–82, 85–88, 90, 98, 116, 176, 202

Quistello 29–30, 34

Reichenbach 124, 126, 142
Religion 84, 99, 107
Riedesel, *Generalmajor* Adolph Friedrich von 48–49, 118–119, 166
Rifles 60, 121, 123, 154
Rossbach 152, 173
Running *see movement, speed of*

Seven Years War 39–40, 51–53, 55, 63–65, 67–69, 78–79, 81–82, 84, 90, 92, 94, 96, 99, 111–113, 115–116, 118, 124–127, 139, 145–146, 149, 152, 161, 163, 167, 169–171, 173, 175–176, 178–180, 203
Skirmishing 19, 22–23, 27, 36, 59, 82, 97, 117, 123–132, 136, 138–139, 165, 176, 192, 203
Staten Island 139, 142
Steuben, Friedrich Wilhelm, Freiherr de 47, 49–51, 58, 88, 123
Sullivan, Major General John 56, 142, 150, 163, 173, 180, 183–188

Torgau 51, 69, 149, 152
Trenton 128–130, 140, 179–181

Uniforms and clothing 39–40, 85–86, 171

Vellinghausen 127, 139

War of Austrian Succession 17, 51–52, 70, 81, 95, 116, 152, 159–161, 178, 180
War of Polish Succession 17, 19, 36, 159, 164, 177
War of Spanish Succession 17–19, 25
Warnery, Charles Emmanuel de 20, 22–27, 33, 36, 105, 126, 145
Washington, General George 50, 57, 80, 91–94, 96, 104, 120, 122, 129, 150–151, 168, 179–181, 183–190, 193
Wolfe, Major General James 57, 89, 92, 118, 155, 159–160
Württemberg-Winnental, Friedrich Ludwig von 19, 22–23, 27, 33, 36

Zorndorf 51, 112–113, 122, 149, 153

From Reason to Revolution – Warfare 1721-1815

http://www.helion.co.uk/series/from-reason-to-revolution-1721-1815.php

The 'From Reason to Revolution' series covers the period of military history 1721–1815, an era in which fortress-based strategy and linear battles gave way to the nation-in-arms and the beginnings of total war.

This era saw the evolution and growth of light troops of all arms, and of increasingly flexible command systems to cope with the growing armies fielded by nations able to mobilise far greater proportions of their manpower than ever before. Many of these developments were fired by the great political upheavals of the era, with revolutions in America and France bringing about social change which in turn fed back into the military sphere as whole nations readied themselves for war. Only in the closing years of the period, as the reactionary powers began to regain the upper hand, did a military synthesis of the best of the old and the new become possible.

The series examines the military and naval history of the period in a greater degree of detail than has hitherto been attempted, and has a very wide brief, with the intention of covering all aspects from the battles, campaigns, logistics, and tactics, to the personalities, armies, uniforms, and equipment.

Submissions

The publishers would be pleased to receive submissions for this series. Please email reasontorevolution@helion.co.uk, or write to Helion & Company Limited, Unit 8 Amherst Business Centre, Budbrooke Road, Warwick, CV34 5WE

You may also be interested in:

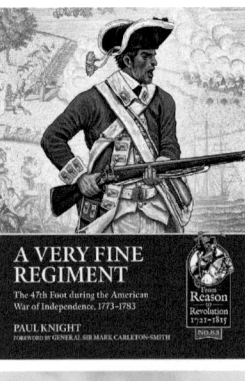

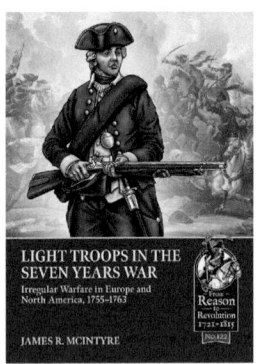
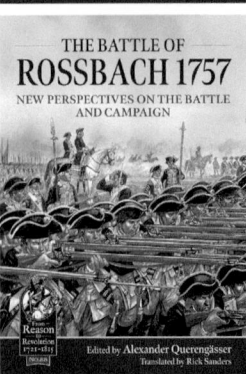
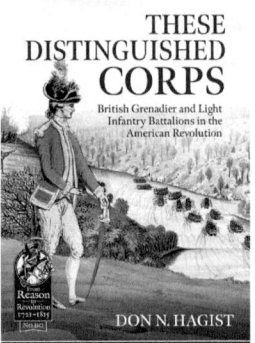
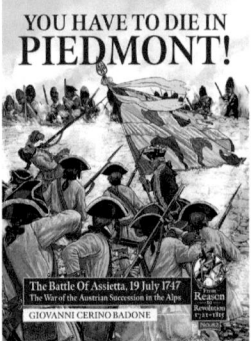